Finding Your Way Through Formal Verification

Copyright © 2018 by Bernard Murphy, Manish Pandey, and Sean Safarpour

All rights reserved. No part of this work covered by the copyright herein may be reproduced, transmitted, stored, or used in any form or by any means graphic, electronic, or mechanical, including but not limited to photocopying, recording, scanning, taping, digitizing, web distribution, information networks, or information storage and retrieval systems, except as permitted under Section 107 or 108 of the 1976 US Copyright Act, without the prior written permission of the publisher.

Published by SemiWiki LLC Danville, CA

Although the authors and publisher have made every effort to ensure the accuracy and completeness of information contained in this book, we assume no responsibility for errors, inaccuracies, omissions, or any inconsistency herein.

First printing: March 2018 Printed in the United States of America

Table of Contents

Table of Contents 2

Foreword 5

Time for a fresh look at formal 9
    Why now? 9
    Why another book on this topic? 11
    Organization of this book 14

The verification treadmill 16
    Why verification is so important 16
    The price of not being perfect 19
    Why is this so hard? 20
    Battling the exponential 24
    Enter formal verification 29

Formal Verification – the Early Years 31
    Background 31
    Equivalence checking 36
    What do we mean by correctness? 38
    Properties: assertions and assumptions 42
    Under the hood: the formal engines 46

What's the catch? 55
    Property checking goes commercial 55
    A problem with assertions 55
    Completeness is expensive 57
    Inconclusives and debug 58
    Stuck at the chasm 61

Crossing the chasm – fearless formal 63
    Pre-packaged solutions 63
    Baby steps – functional linting 65
    Coverage analysis – formal helping simulation 67
    Having the right connections 69
    Sequential equivalence – some assembly required 73
    Other apps 77
    Are we there yet? 79

The role of formal in design today 81
Adoption 81
Size constraints 83
Popular targets for property checking 85
Good targets 85
Targets to avoid 87
Product and service solution providers 88

Adding formal to your flow 91
Organic skills growth 91
First-cut targets for formal verification 95
Detailed planning 96
Monitoring progress 98
Under-constraining and over-constraining 99
Bounded proofs (inconclusives) 101
Manually-guided proofs 103
Getting to formal signoff 104

Looking forward 108
Application domains 108
Functional clock domain crossing analysis 108
Functional power intent verification 110
Architectural formal verification 112
Security verification 115
Safety verification 116
Datapath Validation 117
Technology advances 119
Machine learning 119
Advances in tools and methods 121
The outlook for formal verification 123

If you want to go deeper 125

Glossary of formal terms 126

Acknowledgements 130

About the Authors 132

Foreword

I first became aware of formal verification methods back in the 90's when I would hear presentations at DAC and talk to others who were curious about this area. It looked very interesting, but it always sounded like a future. The problems used in examples were cool, but they were always very small, way too small to face real verification needs, even back then. Mainstream simulation support continued to get better and (more or less) kept up with Moore's law. Working in startups back then, where we had very constrained budgets, formal sounded neat but nowhere near relevant to our day-to-day tasks. Still, I stayed in touch with progress as time allowed.

Around the mid 2000's I joined a large microprocessor company, heading a big verification team facing the usual deluge of challenging verification problems. We had the capacity and budget to experiment with formal and had some success. But progress was in fits and starts. We worked with different vendors and tried different solutions. Some broke, some we were able to make work in some fashion. Still, nothing came out of this that looked like it would quickly go mainstream in our verification flow. Formal was still a side-project.

The big turning point came when we had to figure out how to verify a floating-point unit. We did a lot of interesting work between theorem-proving and property-proving and had great results, which also got other people interested in what could be possible. Progress, but back then formal tool capacities were limited, we could usefully apply them on some blocks, but these were still relatively small.

More recently at Samsung, I had to build a verification team from scratch. We faced crazy schedules (who doesn't in verification), we were looking for ways to accelerate bug convergence and formal was suggested. I was good with this idea given my experience, so we started building unit and system testbenches. It took a while to start finding useful problems (this is something to remember when you are building up formal expertise – you have to be patient). We found a few bugs this way, but payback was still not very exciting.

Once again, a breakthrough came when we found, on the bench, a problem which would occur only intermittently on one machine every 17 days. We had to brainstorm hard about why this might happen and what we eventually concluded seemed impossible. We had a formal setup already available, so we checked it out. Within about 40 minutes we proved that what we thought was impossible was actually possible. This was a really deep bug, about 85 cycles deep. Based on this diagnosis, we worked on confirming it in simulation. Which we were able to do, but this took multiple teams working full-time for 2 weeks to reproduce what formal found in 40 minutes, because it was so complicated to get to the bug.

That made us all believers in formal. We continued to invest. I brought in an industry expert to head the team, we grew the team by adding non-experts, who we coached into becoming sufficiently expert to become effective, and we started to see real success.

As we built up experience, formal bug-finding productivity has shot up. We're no longer finding just a few critical bugs, or a handful of other bugs after weeks of development. On our current project, **20-35% bugs of the bugs found have been discovered by the formal team, and they're finding bugs as fast or faster than the simulation team.**

Another unexpected benefit is impact on better managing shuffling workloads. Teams in big organizations know all about this. Verifications tasks move around as product groups need to accelerate schedules or hiring doesn't move as fast as you expected. We have found that the formal team is a great resource to pick up tasks where simulation teams aren't yet ready to move over. Since formal doesn't need to start with elaborate testbenches, they can often start finding bugs quickly and mature a block along the bug-finding cycle. When a simulation team frees up, they can take the block to completion.

I understand the challenges in adopting formal. You have to invest, and experience shows that takes time to return value. Why take away resource from the certain and well-understood value of simulation to sponsor development around an uncertain future value in formal verification? That intermittent bug made me a believer. Finding it in 40 minutes in a formal property check against many person-weeks in simulation (after formal had already shown what we had to look for) was enough to justify ramping up a formal team. And while finding hard bugs is important, the big bonus is in a complementary verification strength to manage simulation overloads, to accelerate early maturing and to bail out slow programs.

Formal will never replace simulation. All that investment you made in UVM, constrained random, emulation and prototyping will always be important. Formal adds another string to your verification bow. As you build experience, you'll see that some problems you will eventually find in simulation can be found faster with formal. And that's what we're all ultimately after in verification – better coverage, faster and as cost-effectively as we can manage. This book should give you a starting point in understanding how to get to that value.

Jim Greene
Director at Samsung Austin R&D Center

Time for a fresh look at formal[1]

Why now?

You might imagine that the people who build the advanced hardware technologies that you find almost everywhere today would feel comfortable with almost any aspect of technology related to their domain. Or at minimum they wouldn't feel intimidated by any topic. They might not understand it now but, if the need for understanding arises, you expect they would be confident that they can quickly become sufficiently expert, as they have already demonstrated through their mastery of multiple verification techniques: static, directed and constrained-random simulation, along with emulation and prototyping. For rare problems where methods and tools were available but difficult to use, they could always hand a problem "over the wall" to dedicated experts.

It might surprise you to learn that many otherwise expert designers and design managers, if pressed to answer honestly, will admit that they put formal verification in that over-the-wall category and often find it confusing or intimidating. The problem is not so much in broad concepts but in going any deeper, or in knowing how to quantify value. Until relatively recently this wasn't much of a problem. For many, formal verification was at the periphery of the verification toolset. Where a few

---

[1] A quick word about nomenclature: we'll use *formal*, *formal methods* and *formal verification* fairly interchangeably in this book. While this usage is a little loose, it does follow common practice among non-specialists

especially challenging problems defeated conventional verification approaches, they were passed over that wall to experts in formal methods, who would translate reasonable English-language requirements ("we need to check if this can ever happen") to formal tool inputs, then coax the tools into performing their magic and finally provide back either a thumbs-up ("that problem can't happen") or an example of a realistic possible failure.

Valuable though this service was, the impact of formal verification in those early days was limited. Even point problems are important to find, but it was difficult to quantify how this technology contributed to overall verification signoff. Formal methods lacked obvious, much less signoff-quality metrics so signoff (is this design production-ready?) clearly remained the responsibility of traditional verification. If there was interest in using formal methods, executives had to consider the cost of building and maintaining a team of specialists, a worthwhile investment for large enterprises (as we'll see) but beyond the reach of more modest budgets.

How times have changed. Now formal verification stands shoulder to shoulder with simulation, static methods, emulation and prototyping, a co-equal in verification flows across all large and many small design and verification organizations. This is partly thanks to continuing improvements in the capability and usability of tools, but more significantly it has been driven by the relentless increase in complexity of modern designs. Some verification tasks, once solved by throwing more bodies, more licenses, more machines at the problem, have already moved beyond the reach of confident signoff through non-formal methods.

Executives are always concerned about the impact of quality problems escaping to the field; they worry especially about critical components exhibiting intermittent problems from product to product. Could one of these latent problems suddenly explode into a customer crisis? Those same executives are now doubly-sensitized to the media and market fallout that can result from a publicly-exposed safety problem or hack and are actively sponsoring teams and methods to mitigate these risks. Formal verification has become prominent in those efforts.

Why another book on this topic?
There are already many books on formal verification, from academic to application-centric, and from tutorials for beginners to guides for advanced users. Many are excellent for their intended purpose; we recommend a few at the end of this book. But most start from the assumption that you have already committed to becoming a hands-on expert (or in some cases that you already are an expert). We feel that detailed tutorials are not the easiest place to extract the introductory view many of us

are looking for – background, a general idea of how methods work, applications and how formal verification is managed in the overall verification objective.

There are a lot of us who aren't yet at that commitment stage, or who possibly may never want or even need to become hands-on experts. If this describes you, a 300-400-page tutorial may be more than you are ready to attempt; you want something you can read through relatively quickly to get a general understanding of the domain. This book was written for you as a way to dip a toe in formal waters. If you like what you read, you can move on knowing that an investment in serious learning will be worthwhile. If you don't, hopefully you still feel you have gained enough insight to defend, again more knowledgeably, why a deeper understanding of formal methods isn't appropriate to your current objectives.

Who might this describe? You could be:

- *A Designer or Verification Engineer*: You've heard about formal verification, maybe read a little on the topic, or sat in on presentations or tutorials, but you're uncertain whether this direction is right for your needs. You're intrigued by the idea but not quite ready to pick up a textbook; you want to ease into it. This book will get you started with a good broad understanding and should set you up to make that textbook less daunting if that's where you want to go next.

- *A Design or Verification Leader or Manager*: If you're planning to directly manage a formal team, you have to start somewhere. Just like the hands-on engineer, you'd probably appreciate a little orientation before you dive all the way in. Even if you're not directly supervising a formal team, if you're a designer or verification lead or manager, you can expect formal experts to come to you asking questions and looking for guidance about your design, or what is covered in other testing. If such an engineer asks you about an acceptable state-space radius to adequately check a property, you probably would like to know what on earth they are talking about, without having to become a formal expert. We can help.

- **A *Verification Manager or Director***: Here we're talking about people who plan and supervise formal verification activity as a part of their overall verification responsibilities. In verification planning, you certainly need to know where formal can play a role and where it may not be suitable, what effort and expertise should be planned for in using these techniques (like most verification techniques, these generally aren't push-button) and how you can assess effectiveness and coverage in what formal teams report back to you. We aim to help with some insights on getting to signoff with formal verification.

There are others we hope will also find value in this book – those of you who are only peripherally involved in verification or who maybe aren't even in engineering. You might be in applications support in a different domain, in sales or marketing, you might be an executive or even in finance or legal. Perhaps you will never run a formal tool or sit in on a verification meeting, but you feel you could be more effective in your role with a better understanding of this domain. As much as others more directly involved, you deserve (if you have the interest) to better understand formal verification and to see where it fits in enhancing product quality. To serve the needs of this broad audience and in the spirit of an introductory overview, we have kept technical detail to a minimum.

Organization of this book

Since we're writing for a fairly wide audience, we cover some topics that some of you may consider elementary (why verification is hard), some we hope will be of general interest (elementary understanding of the technology) and others that may not immediately interest some readers (setting up a formal verification team). What we intentionally do not cover at all is how to become a hands-on expert.

Chapter 2 presents an overview of the verification problem in SoC design, why this is hard and various techniques common in managing complexity, as an introduction to the role that formal methods can play in the larger verification task.

Chapter 3 reviews the early history of formal verification in our industry, along with some of the basic concepts like assertions and constraints and a very lightweight introduction to the engines that drive formal tools.

Chapter 4 walks through the early challenges formal methods faced in getting to widespread adoption: complexities in setup, running and debug and the level of expertise required to effectively use the tools.

Chapter 5 talks about how formal tool suppliers overcame adoption challenges by introducing apps which provide much simpler use-models for targeted applications. This chapter also reviews a number of the most common apps.

Chapter 6 covers the ways formal is being used today and how wide that usage is. If you want to persuade your manager that an investment in formal is worthwhile, you may find useful evidence here to help build your case.

Chapter 7 is for verification managers (perhaps you) – how can you effectively build and manage a formal verification team and what can you learn from the lessons of others?

Finally, chapter 8 talks about other formal applications you might find useful today or can look forward to in the (somewhat near) future.

We close with a few recommended books/papers you may want to read if you want to dig deeper and a glossary / magic decoder ring to help you with the sometimes-confusing terminology popular in formal circles.

The verification treadmill

Why verification is so important

The goal of verification in semiconductor and system design is to prove that what we plan to build will do everything it is supposed to do and will never do anything it is not supposed to do. This is important in part because the cost to design and build one of these systems now runs to tens or even hundreds of millions of dollars; a trial-and-error approach to getting the design right would take an impossibly long time and become prohibitively expensive.

An even bigger concern for any enterprise is the possibility that customers might discover problems in their products. Issues can arise especially in use-cases that product designers didn't consider and therefore didn't cover in verification. For software, we're all too familiar with patch updates, but issues in hardware can't necessarily be patched; only a new chip can fix the problem. Making field changes to hardware is extremely difficult, in many cases close to impossible. Problems like this can have huge negative impact both on the supplier and their customers.

The ubiquity of electronics

Meantime, the complexity of electronics is accelerating rapidly. We now have high-resolution gaming, smartphones, 4G (and soon 5G) cellular communication, low-power design, cloud computing, semi-autonomous cars, smart homes and the many other high-tech capabilities that surround our modern lives. These System on Chip (SoC) designs are in many ways significantly more complex than earlier systems, in size certainly (thousands of times larger than the Intel Pentium for example[2]), but also in integrating more complex subunits such as multi-core CPUs, GPUs and other complex sub-functions, running multiple different types of software and inter-operating not only with each other but also communicating with the outside world over cellular, Wi-Fi

---

[2] http://www.wagnercg.com/Portals/0/FunStuff/AHistoryofMicroprocessorTransistorCount.pdf

and Bluetooth links. They're also running much faster, with complex dynamic clocking and power management turning functions on and off in the middle of all this activity purely so you only have to recharge your device every few days.

Our tolerance to bugs is dropping. Where once problems in electronics were an inconvenience, fixable in the worst case by a reboot, now advanced systems control safety-critical functions in our cars or pacemakers or power plants. In these contexts, reboots are not an option and failures at minimum may lead to recalls, or worse still may cause fatal accidents. Security has become a major concern. The recently-reported Meltdown and Spectre[3] bugs highlight how far we still have to go in containing security attacks. Where verification must try to find (and fix) every possible way in which a product might be compromised, attackers only have to find one way in and they delight in finding obscure loopholes.

For all these reasons, product teams invest massively in design verification – at least 50% of the total effort that goes into designing a product[4]. Thanks to hard work, clever techniques and continuing advances in verification tools, the industry has released many products which have worked and continue to work extremely well. But as design capabilities and demands continue to race ahead, it is inevitable that these verification strategies,

---

[3] https://meltdownattack.com/

[4] There are differing views on this number – anywhere from around 50% to 70%. One interesting review is http://www.chipdesignmag.com/martins/2008/11/27/the-myths-of-eda-the-70-rule/

comprehensive though they are, have started to show cracks.

The price of not being perfect

The earliest widely-visible instance of a semiconductor verification failure in released products appeared in 1994 when a public post revealed that the (Intel) Pentium floating-point divide returned notably incorrect answers in a very small set of cases[5]. Intel have been verifying complex designs for a long time, they have a worldwide user-base depending on the accuracy of their platforms, they have accumulated massive test suites over years of development, and still a bug slipped through. Design and verification teams around the industry paid attention; if this could happen to Intel, who knew what unseen problems might be lurking in their own production designs?

$$\frac{4{,}195{,}835}{3{,}145{,}727} = 1.333739068902037589$$

The Intel floating point bug - the digits starting with 739 are incorrect

Not finding these problems can be expensive. If you isolate a bug in-house after you manufacture (but haven't yet shipped) the device, you can do more testing but face potentially millions of dollars to fix the design. If the bug gets out into your customer base, costs explode. Intel reportedly took a pre-tax charge of $475M against

---

[5] https://dac.com/blog/post/history-formal-verification-intel

earnings to correct their floating-point divide problem and update customers with the corrected device[6].

Safety and security considerations will further amplify the cost of bugs, perhaps as much in market impact and liability as in replacement costs. Media hair-trigger responses to bad news can drive instant drops in share price and may amplify reputational damage from which it can take years to recover. News of a glitch in an iPad app (used by pilots to access maps of airport runways) drove American Airlines stock down by $1.9 billion in the course of a few hours[7]. This problem was attributed to a software glitch, but we have already seen that hardware is not immune to bugs. Frankly, social media and markets don't care about that hardware/software distinction anyway. The tech failed in a serious way - dump the stock.

Why is this so hard?

At first glance, it might seem that we just need to verify harder or smarter, or maybe both. Unfortunately, no matter what we do, we can never ensure complete verification. It's important to understand why; this starts with how we verify.

The most popular method used in verification is simulation. We create and run (simulate) a series of tests and compare with the results we expect. When running a test returns the expected result, the test passes. When it doesn't there is a discrepancy between the design and the expected result, which may mean we have a bug in the

---

[6] http://www.trnicely.net/pentbug/pentbug.html
[7] https://cdn2.hubspot.net/hubfs/69806/Reassessing_the_Cost_of_Software_Quality.pdf?t=1510935735043

design or it may mean that our expectation was wrong. This approach, simulation-based testing through specific tests, often called **directed testing**, is the natural way we approach verifying almost anything. It's also very effective, so much so that it continues to play a very major role in all verification today.

But it's incomplete. No matter how many tests you create and what clever tricks you use to cover multiple test scenarios in each test, you can only verify correct behavior across a finite number of possibilities, generally much smaller than the total set of possible behaviors.
This is an intrinsic problem in verification for any but the most trivial systems. To make this concrete, imagine the system is a phone and pushing a button on the screen starts a possible sequence of transitions between states which may go through thousands (or millions) of intermediate "next states" before finally delivering the expected outcome, starting a phone call.

Proving that this phone call always works correctly and never works incorrectly should test all of those possible sequences. If there were ten possible options (next states) at each stage and you wanted to exhaustively test a sequence of 100 steps (trivially short for hardware and software these days), you would have to test 10^{100} sequences, a task which would not be remotely possible even on a battalion of supercomputers[8]. The *intrinsic* complexity of verification grows exponentially fast with the number of states in the system (which in a hand-waving way is related to the size of the system) and with the length of the sequences[9].

---

[8] If you could test a billion sequences in a nanosecond, you could test roughly 10^{25} in a year; 10^{100} would take 10^{75} years; roughly a trillion trillion trillion trillion trillion trillion years

[9] Even if you allow only two possibilities at each state, growth is still exponential. It starts slower but still exceeds practical reach very quickly. 2^{100} sequences might be practically checkable but a modest growth in sequence length gets you to 2^{1000}, which is again out of range of reasonable computation power.

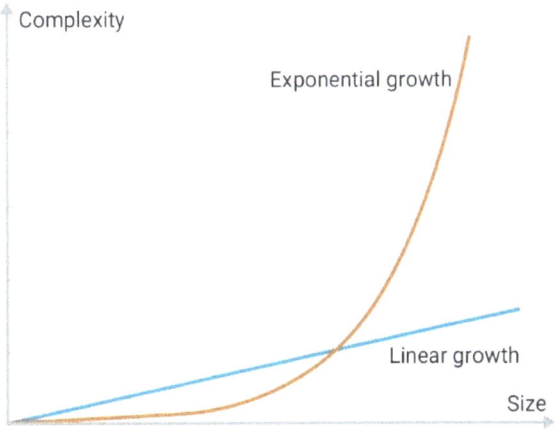

Exponential growth is, with few exceptions[10], the fastest known growth in the natural world, and is much faster than our ability to speed up computers and software (Moore's law notwithstanding). Everyone involved in conceiving, building and testing complex systems works hard to find ways to tame the implications of that growth for verification; clever techniques can often manage the early part of this growth to acceptable levels; as we'll see, formal methods have become important tools in this continuing battle. Still, it is important to remember that because of exponential growth, no one tool or methodology will ever become a long-term silver-bullet solution[11]. Verification will always depend on a range of tools and methodologies.

---

[10] https://en.wikipedia.org/wiki/Hyperbolic_growth
[11] Notwithstanding periodic debates, always entertaining, on formal verification eventually obsoleting simulation

Battling the exponential

Because simulation is incomplete, a widely-employed strategy is to decide when you have run **enough** tests. Verification teams get clever about this by testing bounding or corner cases which lie at the edges of acceptable behavior. For example, when testing arithmetic functions, using biggest possible numbers and smallest possible numbers as inputs is an obvious starting point for bounding tests. The reasoning is that if these extreme cases verify correctly, all other cases in between should also work correctly. This tactic alone can massively reduce the number of tests required for that function.

Unfortunately, corner-case reasoning can be dangerous because it makes assumptions about the way the design is implemented, and those assumptions may not be valid. Often the design team decides that an architecture or implementation must be optimized for performance, size, power or other factors that may not be known to the verification team. In the case of the arithmetic example (say for multiplication) the implementation for small numbers may be quite different than that for larger numbers. Whenever there are architecture or implementation transitions like this, there are new possibilities for errors around those transitions. The verification team now must test not just at the extreme bounding cases but also at these new "architecture boundaries". Corner-case methods continue to be widely used, because we have no choice, but we must always acknowledge that there is an element of human judgement in the corners we pick.

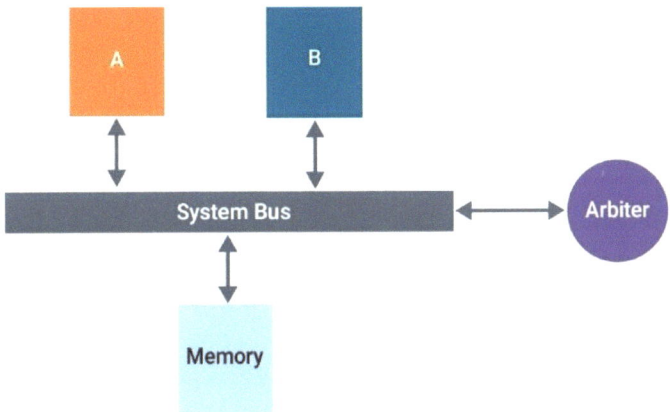

A system bus managing data traffic between 2 IPs and a memory

Another challenge for directed testing is that sometimes the range of possible variations is simply too high for us to even think of all possible options, much less create tests for all those cases. A good example is a traffic manager (called an arbiter) on a bus through which multiple attached components (call them A and B) can communicate to a resource like a memory, but only one at a time .

When A is communicating with the memory, if B also wants to communicate it will make a request but must wait until the arbiter grants it permission. The arbiter may do this after A has completed, or if B has higher priority it may tell A to stop and allow B to start, but must eventually get back to letting A complete its task; it may even allow traffic from A and B to interleave in a controlled way. Since behavior depends on the order and sequence of requests to the arbiter (often from many more connected devices than we show here), relative request priorities, the amount of data to be communicated and what is pending in the request queue, the number of tests needed to prove correct behavior in all possible cases grows quickly[12].

Obviously, building tests to verify that even this basic arbiter works correctly across all possible combinations can be challenging. Adding real-world complications like interrupts, differing clock speeds, variations in architecture for IPs built by different teams, and many more factors, it quickly becomes impractical to build comprehensive suites of directed tests to cover all cases. Even corner-case tactics won't work here – there are simply too many corners.

Facing this problem with directed and corner-case based testing, verification engineers have turned increasingly to a technique known as **constrained-random** testing. In this method, they will build a test allowing for some aspects of the test to be randomized in a controlled way, those controls ensuring that the randomized test behavior remains reasonable. This technique in effect greatly expands the number of tests that can be run. One test-script spawns many tests, which can easily run in parallel by adding server capacity and simulation licenses.

Constrained-random testing has been very successful in teasing out potential hidden errors and is now a major

---

[12] Assume 3 components attached to the bus (with 8 request scenarios – 000, 001, 011, etc), each with 3 possible lengths (short, medium long) and each with 3 possible priorities (1,2,3). If the arbiter should remember up to 5 transaction requests for each function (which may come in any order), you need to consider 8x3x3x5x3 = 1080 possibilities. That's a lot of tests!

component of any strategy in directed testing. But clever though this method is, coverage is still bounded by the number of constrained-random tests that verification engineers can write. While each script spawns many randomized tests around a particular objective, none are clever enough to run all possible tests.

The Synopsys ZeBu emulator

Another response to the scale problem is to use hardware-based acceleration, particularly emulation and FPGA prototyping. These technologies provide huge improvements in performance, running thousands or millions of times faster than simulation, which unquestionably helps a lot. Still, acceleration effectively offers only a constant improvement in performance (big though that constant is).

Yet another approach depends on extensively leveraging proven reusable IP components in designs. If these IPs have been carefully tested and proven on multiple prior designs, the risk that they may exhibit problems in your design should be greatly reduced, at least in principle. This was another big step forward; however, reuse only provides confidence that those components work correctly standalone, as advertised by the vendor. There's no guarantee that **your** design will not introduce bugs in the way it interacts with those IP. Reuse **reduces** but does not **eliminate** the need for testing around those IPs.

All of these techniques are actively used today, but we still always come back to the exponential curve. No matter what we do, testing will never be exhaustive or anywhere near exhaustive because you can never test more than a finite number of sequences. The exponential growth in possibilities to test eventually dominates all of these methods.

So product teams depend on expert verification engineers, designers and managers, who hold frequent reviews, track metrics like testing coverage and bug-trends, and who rely heavily on their experience and gut-feel[13] to decide when they have done "enough" testing. This process is solid enough that the semiconductor industry continues to ship successful products. But those darn designs keep getting bigger and more complicated and a question lingers.

---

[13] It's not just about tools or metrics. Knowing that a particular function used in the current design has had a history of problems in previous products will alert an experienced manager to beef up testing on that function

Beyond all the great testing that has been done, do hidden bugs still remain?

Enter formal verification

A strength of simulation-based[14] verification is that it works naturally with the way we think about testing. We define a test, we write it, we define another test, we write that and so on. We always know how to further expand the range of tests we can supply. But we've seen the limits of this approach, not just in the sense of covering absolutely every possibility, but even in the sense of covering all the important possibilities.

A carefully-designed and executed testplan should cover well all possibilities that we consider important in **normal** use, and also a set of **abnormal** use-cases that we deem possible. But what we consider "abnormal" is based on experience, subjective judgement and frankly, practicality. We have to put a bound on abnormal cases we are prepared to test to be able to complete verification in reasonable time. This can mean that we fail to consider unusual cases where a combination of conditions, building over many cycles, conspires to cause a seemingly impossible behavior, as in the *Meltdown* combination of speculative execution and cache behavior.

A different angle of attack seems to be called for. The problem with simulation is that we must handle testing case-by-case. We can test only at (carefully-chosen) corners, we can cover more testing ground with

---

[14] From here forward we will use "simulation" to cover all the directed (and randomized) testing methods, including emulation and prototyping

constrained-random, we can run many cases in parallel or we can get big speed-ups through hardware acceleration. But all of these methods expand testing capability by fixed amounts; none can overcome the exponential growth of inputs and sequences to be tested. We really need a method that can test all possible cases simultaneously (at least up to some point)[15]. We shouldn't forget also that we want to be able to do this at signoff quality for significant aspects of the verification plan; there's little added-value in any technique which has only incidental impact.

One way to do this is to use variables for inputs and state values and a mathematical model for the design rather than the explicit states used in simulation. To illustrate, think of a 32 -bit integer multiplier. In simulation, we test this works correctly by computing 1x2, 3x5, 7x13 and many other cases. Checking all possibilities requires 2^{64} tests (about 10^{19}) which could take a long time. If instead, we could test with variable ("symbolic") inputs, say A and B, and mathematically verify that the output was always the formula A x B, we could completely prove correctness for all possibilities. This is the objective of formal verification.

---

[15] Perhaps quantum computing can help at some point, though there is no indication such a solution is near.

Formal Verification – the Early Years

Background[16]

The principle behind formal verification is quite simple to state (though somewhat harder to implement) – turn what you want to verify into a mathematical proposition, then prove the correctness of that proposition. This is a very natural direction to take since digital designs are based on (Boolean) logic. You can think of a design as a (typically very complex) set of logical statements, and a behavior you want to verify (maybe "pushing this button will always initiate a phone call") as a mathematical theorem you want to prove in the context of those statements.

> **Premise #1:** All men are mortal
> **Premise #2:** Socrates is a man
> **Proposition:** Socrates is mortal
> **Proof:** Since Socrates is a man and all men are mortal, then Socrates is mortal.

Pythagoras Theorem Proof: $a^2 + b^2 = c^2$

Proof: Total square area is $(a+b)^2$
This area is also $c^2 + 4(1/2*a*b)$
Therefore: $(a+b)^2 = c^2 + 4(1/2*a*b)$
Equivalently: $a^2 + 2ab + b^2 = c^2 + 2ab$
And therefore: $a^2 + b^2 = c^2$

Mathematical proofs

---

[16] A Brief History of Formal Methods

Mathematical proving has a very long and distinguished history, dating back to the earliest Greek philosophers who recognized that it was possible, given an appropriate construction of the problem and requirement, to prove statements which must be universally true. The Socrates example above is a very simple case illustrating the mechanics of a logic proof. You start with **premises** (which in our case correspond to the circuit description and perhaps some constraints on allowed behavior), you assert a **proposition** which you want to prove (in our case an expected behavior of the circuit), then you **prove** that theorem by following a logical and well-grounded sequence of steps.

We'll stress a point here because it underlies the basic advantage of formal methods. When you follow a mathematical approach, and prove a proposition formally, *you have proved it (in a finite number of steps) for all possible cases*. But when you simulate, you only prove for the cases you simulated; if you simulate a thousand cases but there are a million possibilities, you have still only proven a tiny fraction of what the formal method proved. This sounds so good that you might wonder why we still use simulation; it turns out that simulation and formal have complementary strengths (and challenges), as we'll see later.

Unsurprisingly, work in this direction advanced almost exclusively in academia and the big labs, in part because these were interesting technical questions but also because concerns were being raised in the US DoD and telecom companies, among others. Around the 1980s, a general sense that "we need to do better" transformed into more alarmed urgency, prompting active use for formal methods in a number of industries beyond the semiconductor ecosystem.

Boeing Dreamliner

In all cases, adoption of formal methods was prompted by publicly-visible and serious failures on large and critical systems, including a radiation therapy machine delivering fatal overdoses of radiation[17], an Ariane rocket exploding 40 seconds into flight[18], Prius cars with an unexpected stall problem[19] and the recent discovery that the Boeing Dreamliner could lock up and lose control after 248 days of continuous operation[20]. Each failure was ultimately traced to rarely-activated bugs which had been missed

---

[17] https://en.wikipedia.org/wiki/Therac-25
[18] https://around.com/ariane.html
[19] http://articles.latimes.com/2014/feb/12/business/la-fi-prius-recall-20140213
[20] Boeing 787 Dreamliners contain a potentially catastrophic software bug

despite extensive testing yet had or could have had catastrophic consequences.

Just as we saw earlier for semiconductor design, a common conclusion from analysis of these problems was that dynamic verification (simulation) alone was insufficient to deliver high-levels of confidence, especially in safety. Each of these systems providers enthusiastically embraced formal methods with an expectation that they could increase that confidence. In the early days, those companies, along with military, aerospace, telecom and other system builders had to rely on custom-crafted tools adapted from university/lab research. Nevertheless, they proved that formal methods could be effective in proving correct operation, or in finding bugs that might otherwise have been very difficult to track down.

Now formal methods have been used to prove the correctness of driverless operation on one line of the Paris Metro[21] and operations of railway control systems[22]. Toyota applies formal analysis to prove correctness in a variety of car systems[23] and Airbus has been using formal techniques for some time in validating correctness of avionics software[24]. In the very complex world of cloud services, Amazon Web Services (AWS) depends on formal

---

21

https://en.wikipedia.org/wiki/Paris_M%C3%A9tro_Line_14
[22] https://www.prover.com/
[23] Hybrid Systems, Theory and Practice, *Seriously*
[24] Formal Methods for Avionics Software Verification

methods[25] to prove correctness of operation of the various components of their solution.

Paris Metro

Closer to home for readers of this book, Intel took the floating-point problem mentioned in the last chapter as a wake-up call to get serious about formal verification[26]. Just before the turn of the millennium they used formal analysis to validate the Pentium-4[27], reporting that no problems escaped to the field matching the seriousness of the earlier bug. This industrial success, as much as technical advances in tools, contributed to growing interest in the field among semiconductor verification teams.

---

[25] How Amazon Web Services Uses Formal Methods
[26] https://dac.com/blog/post/history-formal-verification-intel
[27] High Level Formal Verification of Next-Generation Microprocessors

Since early formal methods software[28] was developed in academia and labs, only deep experts inside those domains knew how to run these tools. Over time, some of these experts migrated into commercial enterprises (such as Intel) where they started to build wider interest in these strange new techniques. But formal remained a highly-specialized art, barely impacting production design flows except in one immediately useful application requiring very little understanding of the underlying technology – logic equivalence checking.

Equivalence checking

When logic synthesis from RTL started to take off, an obvious question arose: how do I know the synthesis tool didn't make mistakes in converting from RTL to gates? When design sizes were relatively small, signoff verification[29] (mostly through simulation) was still common at gate-level so equivalence between the RTL and the gate-level implementation wasn't a primary concern – all that mattered was that the gate-level implementation behaved correctly. But as design sizes grew, high-coverage gate-level simulations became impractical; signoff verification increasingly moved to RTL, making the question of functional equivalence between the gate-level implementation and the RTL a much more pressing concern.

---

[28] For example, SMV and ACL2
[29] **Functional** signoff verification, just to be clear

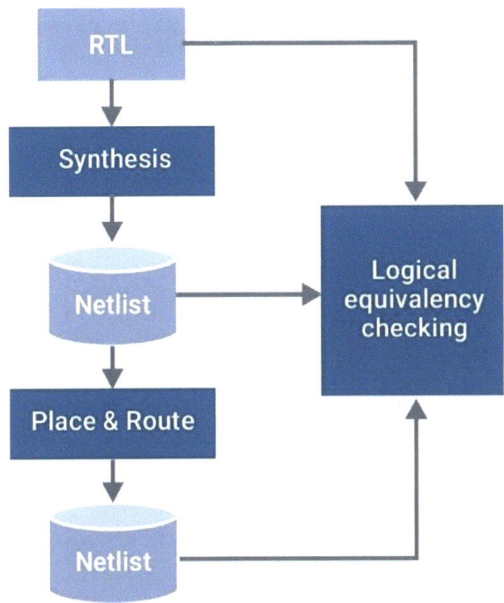

Addressing this need presents a perfect opportunity to apply formal methods. Any formal proof requires some kind of reference against which you're going to check. In this application, we have a ready-made reference – the RTL. We want to check that the **gate-level implementation** functionally matches the **RTL design**. There's no need to create additional statements about what should be checked, which simplifies usability[30]. Thanks to this ease of use and completeness in proving, logic equivalence checking has become a required signoff step in all major production design flows.

Useful though equivalence checking application is, it still doesn't prove "correctness" in the more general sense we probably would like to see – correctness against intended

---

[30] In practice, in modern flows dealing with complex logic, equivalence signoff is still not completely pushbutton

behaviors of the design. Sure, the gate-level netlist matches the RTL, but how do we know the RTL is correct? Or that the architecture is correct? This requires a different kind of analysis.

What do we mean by correctness?

We all believe we know what it means for something to be correct, but correctness is one of those attributes that's not easy to define precisely. We tend to think the same way as Supreme Court Justice Potter Stewart who, when arguing about a definition of obscenity said that "I know it when I see it". This may work in legal decisions but is not useful for the kind of proofs we need.

US Supreme Court

When we take a mathematical approach, we need a precise specification of correctness. Here we run into a problem: what can we use as that specification? Perhaps we could start with the documented requirements? Every design, IP and block (at least for significant components) has some kind of specification, perhaps in Word or PDF. But these are written in natural language (perhaps English) which rarely rises to mathematical precision:

> **English Statement:** Every access request is granted.
> **Possible meanings:**
> 1. Every access request is immediately granted
> 2. Every access request is immediately granted at the next positive edge of the clock
> 3. Every access request is immediately granted at the next positive edge of the clock, if the block is not in reset mode
> 4. Every access request is eventually granted
> 5. Every access request is granted after some allotted time
> 6. Every access request must be paired with a grant

Why English "specifications" struggle with precision

If documentation specifications won't work, perhaps we can use an architectural or SystemC or C model? Models of this type are sometimes available, but they are typically developed to explore and validate high-level features of the design; they are not normally defined with sufficient precision to act as references against which RTL equivalence can be checked. Think of a specification for an integer multiplier: "c = a * b". This may be sufficient for architectural modeling, but it doesn't specify timing (how many cycles are required to complete a multiplication) or power intent (how or when clocks or power should be gated), among other important factors.

In the few cases where these high-level models can be used as a reference (or extended to become an effective reference), then high-level equivalence-checking tools can verify the equivalency of the two models. In practice today, these opportunities tend to be relatively rare and require significant investment in setting up and proving[31].

More commonly, whether starting from a natural language specification or an architectural model, design teams find that so much detail must be added to make the description sufficiently precise for formal proving that the effort required to build this reference outweighs the benefit[32].

> **Larger Ambitions**
>
> There is very active research in an area commonly called theorem-proving in which more extensive statements about what is expected of a system are proved through engines like ACL2, TLAPS, Coq, Isabelle and others. You might think of this as a more powerful use of property-checking requiring expert use, which can generate very powerful proofs.
>
> An unavoidable challenge in using this approach is that it requires a very detailed understating and description of what constitutes correct behavior. Such a description may be at least as complex as the design RTL, both in intent and also in mathematical representation

---

[31] Another approach is to develop a high-level specification to be used for checking in a language such as TLA+. Both AWS and Microsoft Azure teams use this method in testing their software. Obviously, this approach requires further investment to learn and to build specifications and still requires an equivalence checking step for completeness (which is often skipped due to complexity)

[32] This is inevitably a balance between effort and economics or safety, which is why some verification efforts make the investment in comprehensive proving for some aspects of designs

A more practical approach to building a reference specification is to limit the scope of what we are trying to prove to local expectations of correctness, since these can be much easier to describe. Think again of our earlier bus-arbiter example; generating a full specification for such an arbiter could be extremely complicated, especially for complex bus protocols like AXI. Instead you might in some cases choose to use simulation for some of your testing and formal for a **subset of behaviors** that are otherwise very difficult to test.

As one example, the arbiter must communicate with blocks through request and grant signals and will need to store pending requests in a FIFO. A very important specification is that this FIFO should never overflow, because if it does, requests/grants will be lost. Testing for this possibility could be very challenging in simulation because you have to create heavy traffic/demand on the bus to overflow the FIFO. Even then you couldn't be sure that there might not be a case which would create such a problem among the many other possible configurations of traffic and demand that you hadn't tested.

A formal check for FIFO overflows addresses this concern; you can prove this specific problem can never happen (or isolate a case where it could happen). This approach, working with specifications which target important requirements within a function, is used in industrial flows today and is known as ***model-checking*** or ***property-checking***.

In certain cases, it is very possible to accumulate sets of property checks to provide a complete specification for a function, in which case formal verification can assume complete responsibility for the verifying the correctness of that function – no simulation required. This method is known as end-to-end (E2E) checking and is generally considered to be a fairly advanced use of formal.

Recapping, in common usage we abandoned hope of proving correctness against a complete specification and are now limiting ourselves to localized proofs of correctness, where correctness is expressed through properties, as we'll see next. In practice this is not a significant compromise; a complete specification may be redundant if simulation already provides much of the necessary confidence in the correct working of the function. Formal verification then complements this testing with targeted confidence for some especially challenging cases.

Properties: assertions and assumptions

Now we know we are checking properties, what are these properties? We'll start with assertions, which will lead us to properties. An assertion is just what it sounds like: "I assert that these two inputs can never be active at the same time", a behavior on which you depend but which may or may not be true in practice and which you therefore need to check.

The concept and practice of assertions was originally conceived[33] as a way to check *basic expectations* in software through executable checks embedded in the code; these would trigger automatically if requirements expressed through assertions were not met. An obvious

---

[33] As early as the 1940's by Alan Turing
http://www.turingarchive.org/viewer/?id=462&title=01

example before a division operation would be an assertion that the divisor is not equal to zero, since division by zero is not defined. The intent is to catch basic problems quickly before they lead to later and more complex bad behaviors which might be more difficult to debug.

Assertions have been supported in hardware description languages for quite a long time, but widespread use of assertions as a part of hardware verification, known as assertion-based verification (ABV)[34], became popular in the early 2000s after standards emerged. From there, these evolved through OVL and PSL to the leading SVA (SystemVerilog Assertions[35]) standard of today. ABV continues to be very useful and popular in simulation; most importantly, for our purposes, formal tools adapted to read this same standard format[36].

In hardware design today, assertions are predominantly expressed in the SVA format; these can be embedded in the RTL for a design or can be provided through separate files. A simple assertion would be ***assert A == B***[37] which checks that signal A is ***always*** equal to signal B. If a formal tool proves this assertion, then the statement is true in all cases; conversely, if the tool finds this assertion is not correct, even for a single case, then it will report a case it

---

[34] http://www.ijmetmr.com/oljuly2015/NKarthik-MGurunadhaBabu-MuniPraveenaRela-113.pdf
[35] SVA is a subset of the SystemVerilog standard
[36] Earlier, formal tools used languages like CTL, LTL and Sugar for property specification
[37] The actual SVA assertion is slightly more complex. We'll stick to this simplified form here

has found where A is not equal to B (this is called a *counter-example* or CEX).

OK, those are assertions but what are properties? A property is a formalized statement about the design with no attached expectations. For example, "my car drives forward" is a property I can associate with my car. It doesn't imply I can drive forward; that requires an assertion on the property. If I **assert** that "my car moves forward", now I am making a statement that it should move forward. If a property-checking tool could check this, it might report that "yes indeed, your car can move forward" or it might report that "no your car cannot move forward because it's out of gas". In the example in the previous paragraph, **A == B** is a **property** and **assert A == B** is an **assertion**. An assertion checks the property following the **assert** keyword. This division of terms isn't nit-picking because properties can also be used in other contexts, as we'll see next[38].

---

[38] In casual/common usage, *property* is often used as a synonym for *assertion*. Even the experts do this!

Sometimes it is necessary, as a part of proving an assertion, to constrain certain signals so that unreasonable or uninteresting possibilities are not considered. This can be done by defining **constraints**; which look very similar to an assertion except that the keyword is **assume**. Using the same property, A == B, in this case **assume A == B** is an example constraint[39]. But where the corresponding assertion checks for cases where A is not equal to B, the constraint limits checking to just those cases where A is always equal to B.

In our car example, where the assertion is "does my car move forward?" a possible assumption could be "assume my car has gas". Together, the problem could be expressed as "assuming that my car has gas, can it move forward?" With this constraint, the formal tool could come back with "yes, the car can move forward" or maybe "no, it cannot move forward because the gearshift is broken" (isolating a more serious problem).

That's what assertions and constraints do; what do they really look like? Our goal isn't to help you write or even understand the detail behind properties, but it's worth knowing how to recognize the real thing in the wild. They can look rather complicated, but only the hands-on experts need to understand this stuff in detail. And actually, they really aren't as complicated as they first appear. Still, we don't want to scare you off, so we

---

[39] Yes, this is a little confusing; is it an assumption or a constraint? When speaking about them, you can use either term, but **constraint** is most common. But the standard uses **assume**. Sorry, that's just the way it is.

promise this is the only place you'll see these formats in this book.

> **Check that bus is a one-hot encoded signal (only one bit is high at a time)**
> bus_onehot : SVA: assert property (@posedge clk) disable iff(reset) $onehot(bus));
>
> **Check that when a request comes, an acknowledge should come within 5 cycles**
> SVA: req_ack_in5 : assert property ((@posedge clk) disable iff(reset) $rose(req)) |-> ##[1:5] $rose(ack));
>
> **Constraint that there request cannot remain high for more than 5 cycles**
> SVA: assume property ((@posedge clk) disable iff(reset) $rose(req) |-> ##[1:5] $fell(req));

Examples of SVA assertions

Now we know how to describe what to check (through assertions), we have to dive a little under the hood to understand how these properties/assertions are checked. We'll promise not to get too technical here.

Under the hood: the formal engines

At the core of any formal verification tool, you're going to find the model-checker. **Model-checker**[40] is just a fancy name for an engine that will take a circuit and a property (or rather an assertion) and will determine if that property will hold true in that circuit in all possible cases. If that doesn't hold, it should report a failing case, appropriately called a **counter-example** (CEX).

---

[40] Ed Clarke pioneered model checking especially while at CMU and has received multiple awards for his work in this area.

To help understanding, we'll use a simple example – a familiar traffic light controller problem. In this case, we have two cross-streets with lights in both directions at the intersection. Lights in each direction can cycle through red, yellow and green states. The obvious safety property[41] we want to check is that we can't get green in both directions at the same time. If this should fail to happen under any circumstance, the outcome could be catastrophic. Checking this requirement is an excellent application for formal.

Think of a model for this controller based on two finite-state machines (FSM), one for each direction, describing when and how the lights can change. These FSMs need to negotiate to determine which set of lights is going to change to what state (red, yellow or green) next. Each FSM has to consider not only its own current state and next possible state given various inputs (e.g. is there a car

---

[41] A safety property is a property which must always hold true, popularly summarized as "nothing bad happens".

stopped at my light) but also the current state of the other FSM. You should see now that it is tricky to know for certain that the safety property will never be violated, especially when there may be yet more inputs like pedestrian crossing requests. This type of property-checking is also very relevant in SoC bus design where only two devices can be allowed to communicate through a common bus at any time.

Traffic lights also provide a good example for liveness checking[42]. A light might avoid violating safety checks by never turning green, but this is also not desirable behavior, at least for drivers stuck on red. We need to add a property check that each light will turn green eventually (within some acceptable limit in practice). Similar conditions apply in SoCs, again especially around bus communication. An IP wanting to communicate through the bus should not be stalled indefinitely (often associated with hangs or deadlocks)[43]. Each should get a chance to communicate no matter what other demands there might be on the bus.

---

[42] A liveness property is a property which should eventually be true, popularly summarized as "eventually something good happens"

[43] There are very real problems SoC designers watch out for in these cases: **deadlock** where control is stuck in one state and nothing happens, **starvation** where one resource is blocked from access while others continue to have access and **livelock** where two or more resources are locked in a struggle for control and still nothing useful happens!

We don't need to discuss here how the controller is designed, only how we are going to check those properties we specify. Remember that the simulation approach to exhaustively verify the design would be to cycle through all possible input and state combinations, and to check that no assertions fail. In a formal approach, instead we will calculate with variables in place of those explicit values and we will use mathematical techniques to reach a conclusive proof. Here rather than simulating, we build equations expressed in Boolean logic form; these equations cover all possible values, so if we can prove our properties must be true given this set of equations, we have proved it in all possible cases. This approach, called **Model Checking**, was the first big step towards property checking in hardware verification.

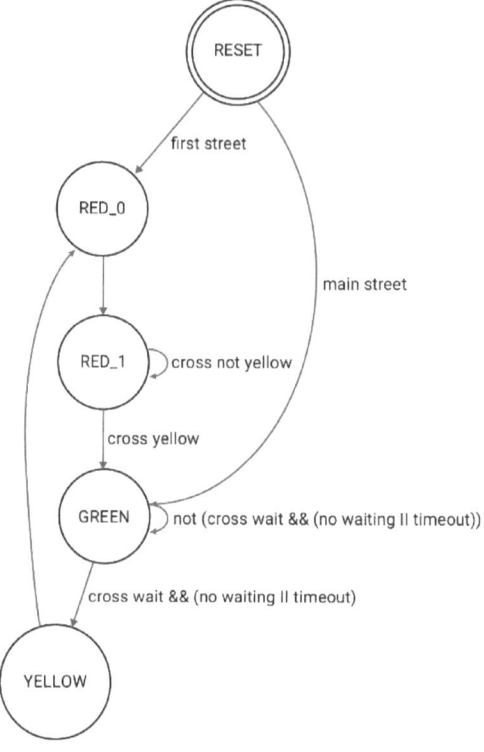

An example finite-state machine (FSM)

The details of how this is done are too technical for our purposes, so we'll attempt a ***very*** simplified explanation. Any set of Boolean equations (and therefore any digital logic circuit) can be represented as one or more interacting FSMs, which can be graphed as a set of states (the bubbles in the picture above) and possible transitions between those states (the arrows). Our traffic-light controller can be graphed in this way. This graph can in turn be mapped different types of graph which are designed to be more efficient for proving properties.

The flow to accomplish all of this starts with the front-end of a logic synthesis flow, which maps your RTL into a control/dataflow graph, from which it builds and optimizes one of those specialized graphs. This correspondence between logic synthesis and formal verification shouldn't be too surprising; the Berkeley ABC[44] platform is a widely used and adapted platform of this type, expressly designed to serve the needs of both synthesis and verification.

The optimized graphs come in different flavors, depending on the engine. One style is the binary decision diagram[45] (BDD). Model checking based on BDDs, called Symbolic Model Checking (SMC), became the forerunner of all modern model checkers and made possible property-proving for designs with 100-200 flip-flops. Impressive progress, but hardly up to the needs of modern designs or even sub-functions. Naturally this triggered more research, to the point that BDD-based SMC methods can now handle designs up to thousands of flip-flops (if components like memories are abstracted out in some manner).

 While SMC with BDD showed promise, capacity was still a major concern. BDD memory consumption grows exponentially with the number of states in the circuit, spectacularly shooting to sizes over 4GB in a matter of seconds, making the approach impractical for many real problems. Looking for a different approach that didn't so quickly succumb to unusable growth led researchers to

---

[44] http://people.eecs.berkeley.edu/~alanmi/abc/
[45] https://www.cs.cmu.edu/~emc/15414-f12/lecture/bdd.pdf

Bounded Model Checking (BMC), using a proof method called **Boolean Satisfiability Solvers, or SAT Solvers**[46].

```
X = b OR c
y = !a OR !d
z = !b OR d
p = !(x AND y AND z)

One SATisfying assignment for p to be
false (0) is:
        a=0, b=1, c=0, d=1
```

An example of SAT on a simple logic design
SAT tests if property p can ever be false; in this case it can

Instead of building the complete problem representation required in BDDs, BMC+SAT switched to a breadth-first approach, looking for a violation of the property to be checked (a counter-example) within a pre-determined bound on clock-cycle depth.

Bounding the depth to which the search extends can significantly reduce the size of the analysis problem, making proofs (and especially finding counter-examples) much more feasible both in memory requirements and in time. At the same time, BMC naturally handled sequential behavior by **unrolling** sequences. The next cycle in logic (for all possibilities beyond this cycle) is expanded as a new

---

[46] SAT techniques have been around for a long time, getting their start in artificial intelligence for applications in planning / scheduling

set of logic, taking the previous cycle states as inputs. And so on for continuing cycles, out to whatever bound is set. Analysis can then just work with this sequence of logic stages without having to worry about clock cycles.

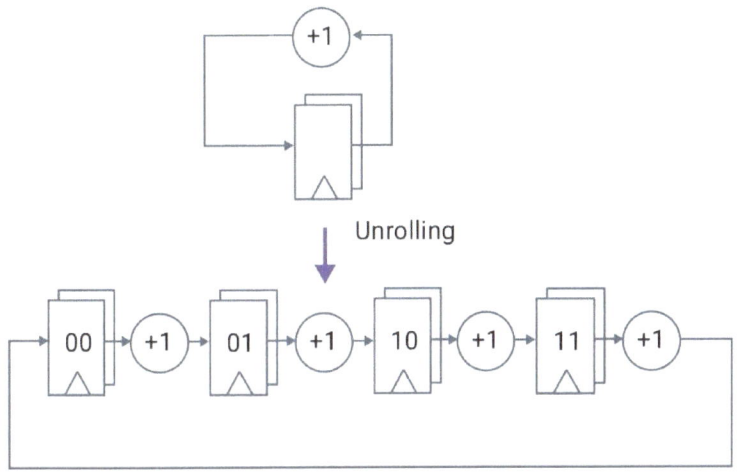

An unrolling operation on a 2-bit counter

The boundedness of BMC working together with the natural solution-finding power of SAT has proven very successful, so much so that now BMC+SAT is now one of the dominant approaches to model-checking. But it's not perfect; while bounded methods will often satisfactorily conclude proof of a property, they can in some cases fail to prove that property or fail to find a counter-example within the bound that has been set. In this case, the result is **_inconclusive_**. We'll talk about this more in the next chapter. Inconclusives (also known as bounded or undetermined proofs) are an unavoidable feature of all formal methods but, as we'll see later, they don't have to be dead-ends.

Tools and methods have continued to evolve at a rapid pace so now there is quite a range of engines, techniques and flows to formally attack a property-checking problem. Exploring all of these would take us too far from our goal of providing an introduction to the field. If you want to learn more about this rich set of possibilities, check out our suggested reading list at the end of this book. And remember that innovation in formal methods hasn't stopped; you should expect to see yet more capabilities appearing in production tools[47][48].

Among this range of proving engines, each engine has strengths in addressing certain problems and weaknesses when facing others. For this reason, we need access to a toolkit of methods to attack the wide range of problems that will arise in real circuits. Managing our way through these options is a topic for our next chapter.

---

[47] http://fmv.jku.at/papers/prasadbieregupta-sttt-7-2-2005.pdf
[48] http://www.springer.com/us/book/9780387691664

What's the catch?

Property checking goes commercial

Thanks to the promise of property checking and success in some high-profile design companies, several commercial products started to appear around the early 2000's, some from the larger EDA vendors, others from new ventures, and were actively promoted as a new direction in verification.

In each case, leading edge verification teams were enthusiastic, using these tools primarily to address the hardest problems that were proving intractable for conventional verification. Naturally, everyone assumed that over time more verification teams would become comfortable with formal methods and adoption would quickly spread. Some enthusiasts even hoped that formal verification might eventually replace simulation. But it didn't work out that way. Why it didn't might be attributed to several factors and is the subject of this chapter.

A problem with assertions

The idea of adding assertions to a design is simple; the practice of adding effective assertions is not always so simple. The easy cases (a queue should never overflow for example) represent a small fraction of the cases you really ought to check. And putting lots of easy assertions throughout the design isn't generally very useful. More important usually is to check more complex bounding cases dependent on multiple states and tricky sequences, one common example being checking for correct interface behavior between blocks in the design.

In writing the associated assertions (and assumptions), you have to think hard about the theoretical operating bounds of the design in order to correctly draw the line between legal and illegal operation. Draw this line in the wrong place and you may report errors on operations which are legal, or you may fail to error on operations which are truly incorrect. Getting this right can be quite tricky since you have to imagine the limits of legal use cases, whereas in simulation you just run tests to see if any bugs appear[49].

Many verification teams found the investment they had to put into thinking of, creating and debugging high-value assertions for formal verification was sufficiently onerous[50] that they would build some, but overall assertion density (as a measure of how effectively you were using assertion-based verification) was not very high. Highlighting this problem, checking the correct behavior of an interface IP (requiring assertions running to hundreds of lines) would have been far out of reach for a typical verification team. And where teams had already adopted assertion-based verification (ABV) for simulation methodology, they often reported that adding constraints ("assumes") to correctly bound proofs took as much effort (or more) than developing the assertions.

---

[49] OK – simulation teams work hard to find those cases too. But absolute proofs should be the central value of formal, so "best efforts" don't really measure up

[50] The effort required per assertion (or group of related assertions) isn't abnormally high. It is often on the same order as building UVM stimulus generators and monitors/checkers. But it's added effort which must be traded off against other verification investments

Completeness is expensive

Remember that formal methods work by analyzing a symbolic model of the logic, rather like solving algebraic equations generally. This has the great advantage that proofs (or bugs), when found, are certain; there's no need to test additional cases. But you pay a price for this completeness. Problem complexity in formal methods grows exponentially with circuit size, even with the most powerful formal engines; in fact, there is no guarantee that any given problem can be solved in a finite run-time on any size machine[51].

This is not the only problem in design engineering which is theoretically unbounded. Place and route (particularly routing) is an instance of the travelling salesman problem[52] which also theoretically may never complete in reasonable time/space bounds. Yet place and route is absolutely routine in digital design today. Formal methods have a similar limitation but still continue to be useful in finding difficult bugs beyond the reach of simulation-based methods. In both these applications, what could in principle be impossible has been wrestled into practical usefulness in most cases through significant advances in empirically-discovered best-practices. But we should remember that completion is not guaranteed; some cases may still require impractical or even unbounded run-times or memory.

---

[51] Famously proved by Alan Turing in 1937: https://en.wikipedia.org/wiki/Turing%27s_proof
[52] https://en.wikipedia.org/wiki/Travelling_salesman_problem

Since formal methods are bounded, if the problem space becomes too big you could just surrender. But sometimes it is worth switching to a different formal method because maybe that proof or bug you are looking for is just a little further out. Remember all those different engines and techniques we talked about in the last chapter? Tricks to see if it might be possible to complete a proof are to try a different engine or to try the same engine with different parameters. Or you might try decomposing the problem into smaller pieces which may be easier to solve separately. In fact, multiple techniques can be applied. Managing all of these options starts to require more expertise on the part of the verifier, which becomes more apparent in the next section.

Inconclusives and debug

If during proving the formal tool stops, it can report one of three possibilities for each property: that a property has been proven, or that a counter-example has been found (maybe a real bug or an artefact of insufficiently-considered constraints), or in some cases that the run was unable to run to completion beyond an acceptable bound. Outcomes of the last type are known as ***inconclusive*** and happen when the tool exceeds a set bound in memory or time. Then you have to consider your options. A number of possibilities were mentioned above – use a different engine or change parameters for the engine. Another option is to manually guide the flow of proving, though various methods.

A common way to reduce the size of proofs is to replace an embedded block of functionality with a simpler model covering only what you believe to be the most important behaviors, a process known as **abstraction**[53]. This may be as simple as replacing a block with a black-box, if that functionality isn't important to what you want to prove. Going back to our earlier car example, if we want to prove that a car can move forward, we don't need to worry about the details of windows, windshield wipers, infotainment and so on. We can start with an abstracted model of the car with only the engine, wheels and drivetrain. We can't model the transmission as a black-box, but we might abstract to a simpler model, considering only the *park* and *drive* states and ignoring *neutral*, *low* and *reverse* options.

Abstracted car

In the design world, we might model a memory as a black-box, effectively allowing for any possibilities and sequences in address and data behavior[54]. Conversely, counters can be challenging for provers because these

---

[53] The Art of Abstraction
[54] Which might be an OK choice in some cases and not OK in other cases – depending on use-model

have also many possible states, but a black-box model may be too unrealistic. Instead, you'll typically abstract the counter with a greatly-simplified model which maybe considers only values significant to downstream logic. For example, if a downstream FSM changes at counter values 2, 5 and 10, only these three values may need be modeled in the counter; all other possibilities are collapsed into a default case. Proving then has a much smaller state space to handle and has a higher chance of completion. But doing this correctly isn't trivial; you have to reduce the state space enough to enable completion but not so much that you may miss real problems.

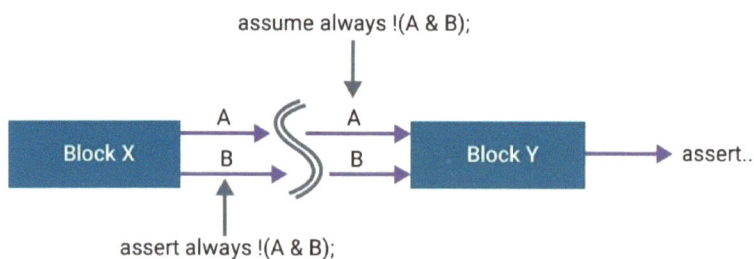

Assume guarantee

Another approach is to decompose the problem into smaller parts as shown here and use a technique called ***assume-guarantee*** at the interface between those blocks. Since each part is smaller than the original problem, property checking in those independent parts is more likely to complete successfully. The blocks are connected through properties which are used as constraints (***assumptions***) on the inputs to the downstream block. Those assumptions are in turn verified (***guaranteed***) to hold at the outputs of the upstream block. Careful use of this technique can reduce problems which are unsolvable, or which complete only in hours, to a set of sub-problems each of which can be proven in minutes. But of course, you have to figure out how best to divide the problem and what invariant properties you will use at each interface. You also need to make sure the way you divide doesn't generate false problems. All manageable, but this is not for beginners.

There's yet another approach to manage difficult proving problems, using constraints. These will limit the scope of a proof by forcing certain signals to take a limited range of values. For example, a USB IP may be configurable to run in 32-bit mode or 64-bit mode. Either mode is legal, but the IP may only be used (at any one time) in one of these modes. A formal tool won't figure this out on its own; you have to specify a constraint. If you don't, it is quite possible that you will get an inconclusive result or perhaps a meaningless counter-example reflecting unrealistic usage. This technique, systematically splitting a problem into separate use-cases is known as **case-splitting**. Often the cases are fairly obvious, but you have to be certain or somehow prove (perhaps using assume-guarantee methods) that there is no possible interaction between operation in different cases.

Stuck at the chasm

You might now have a sense of why formal verification didn't instantly spread everywhere. Where it works, it works very well. But in many cases, getting it there can take quite a lot of expert supervision and effort. Those experts were able to figure out which proof engines to use with what parameters when something got stuck. And they knew how to apply appropriate guidance to the tool to confidently validate behavior without hiding problems.

Design teams and tool vendors quickly learned that the most successful way to deploy formal tools was to build an army of formal experts and farm out all the formal problems to them. Some of the tool companies followed suit, building teams of highly expert AEs, many with advanced degrees in formal verification, who would work closely with customers, in many cases running the tools for them. The need for help also prompted new companies who specialized in consulting for formal applications.

This service-intensive approach worked but necessarily limited scaling use of the technology. Formal verification couldn't expand to being used widely because there weren't enough experts available, and even if more experts could be trained, that service-based use-model would be too expensive to scale in the long-term. In the terminology of Geoffrey Moore's book **Crossing the Chasm**[55], formal verification was stuck on the left side of the chasm. The experts on the left side (small teams in perhaps ten large companies) were happily using formal in expert use-models, but there was no way this kind of usage could cross over to the larger market and mass adoption. Something had to change.

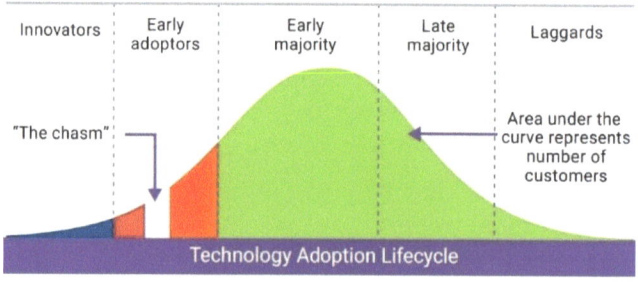

Crossing the Chasm

---

[55] https://en.wikipedia.org/wiki/Crossing_the_Chasm

Crossing the chasm – fearless formal

Pre-packaged solutions

To recap, sometimes formal methods would find critical problems, but sometimes they wouldn't, or couldn't deliver a useful result without additional complex effort. The return on investment was uncertain for many verification managers, indeed even for the companies supplying these tools.

Pre-packaged solutions seemed like an obvious answer – we have verification IP (VIP)[56] for simulation, why not also for formal? Assertion IP (AIP) (also known as assertion-based VIP or AB-VIP) are indeed a part of the answer and vendors offer solutions for a range of interfaces. The same concept is also scaled down to other simpler yet still widely-used design components such as FIFOs, linked-lists and special CDC synchronizers, where assertions/constraints can be packaged with that design IP.

Over time, formal experts found a complementary approach in what is now often known as design or

---

[56] A verification IP is a model used in place of a design IP (such as a USB function) to verify correct interaction of the rest of an SoC design with this IP. VIP are heavily tested against associated standards and are widely used for their reference quality. Simulation VIP also include verification support through assertions, debug support and cover properties

verification *patterns*. These aren't associated with blocks in a design necessarily but rather with commonly-occurring verification objectives and processes. Each time the formal team would address a certain type of verification objective, say checking clock connectivity at the SoC top-level, they found they were building similar scripts, similar assertions and constraints and applying similar abstractions, even following similar paths in decomposing large problems.

Whenever patterns emerge, that's a strong hint that it should be possible to provide value in a different type of automation – an *Application* or *App*. Pre-canned scripts, assertions, etc. aren't sufficient to handle many possible SoCs with different architectures and objectives, but a combination of a greatly simplified user input along with application-specific code can, under the hood, construct and drive all the steps in a specific pattern. This includes not only problem setup but also running and managing run-time issues through all the methods we described earlier. You'll even find app-specific debug in many cases. The app approach really caused formal adoption to take off, so much so that today you'll find around 10 apps offered with each of the major formal platforms.

Typical Formal Apps

A possible misconception about apps is to view them as the beginner's version of formal, to be abandoned as soon as you have built enough expertise. That view is not correct. While apps simplify use of formal methods in their target objectives, they are not verification lightweights. Even advanced verification teams continue to address high-value problems through apps, much more effectively than they could through other verification flows. In fact, it is not unreasonable to expect that over time more standard patterns will emerge and be handled through yet more apps. It is arguably better to consider Apps (and AIP) as the backbone of formal application, with custom property verification reserved for those cases not yet covered by packaged solutions.

Since most verification (and some design) teams get their start in formal through these apps, we'll discuss a few of these in some detail. To avoid confusing generalizations across different products we'll use VC Formal and a few of its associated apps, in rough order of required user involvement, to describe applications. You should remember that products and apps from other vendors may differ in some features and/or use models.

Baby steps – functional linting
The simplest app goes by a number of names – auto-extracted properties (AEP in VC Formal) or functional lint. This app looks at the RTL for a module or block, generating a number of assertions representing standard "best design practices" which are then checked automatically. You never have to be concerned with the internals of those properties. Some of these checks are often associated with linting, but the formal versions checked in the app are less "noisy" (report less false errors) than you would find in a pure lint check. Most important for those who want to get

started with formal, running these checks is almost[57] as simple as running linting. As a bonus, when an issue is found, it is accompanied by a waveform, so it is easy to understand the problem.

```
reg [7:0] a,b,c;

...
assign a = a & 63;
assign b = b & 127;
assign c = a + b;   // a typical linter would flag this as an overflow hazard
```

*Apparent Lint problem
which is not a real problem in this case*

Take for example an arithmetic overflow check. Suppose the RTL code adds two eight-bit (unsigned) numbers and puts the result into an eight-bit (unsigned) register. It is possible this value can overflow the result, which is what a lint-check would report. If you add 16 (decimal) to 240 (decimal), the result is 256 which requires 9 bits and therefore overflows the 8-bit result. However, if the first number is in practice limited to never be bigger than 63 and the second number is similarly limited to never be bigger than 127, no error should be reported. Proofs in this case require formal analysis to determine the functionally-possible bound on the total sum. Here, the formal check will be less noisy than the lint check, minimizing engineering effort to check false errors which makes this app popular in design teams.

---

[57] You may need to add some constraints in some cases

There are other checks in the AEP app, for example checking that a high-impedance bus can never have more than one active driver and that it can never be floating, or that a finite-state machine (FSM) has no inaccessible states (states that cannot be reached through transitions from other states in the FSM). The main point about all of these checks is that they are really as simple to run as lint, and any counter-examples (violations) will be reported in the standard verification debug window. This starting step into formal is so easy that anyone who understand RTL can use it, without any understanding of formal verification.

Coverage analysis – formal helping simulation

The later stages of coverage-driven signoff are always painful. As you build and run test-cases, initially coverage rises steeply. But the more you progress, the harder it becomes to increase coverage. You keep adding more tests after carefully studying which parts of the design are not yet being hit in testing, yet each new test barely moves the coverage needle, if it moves at all.

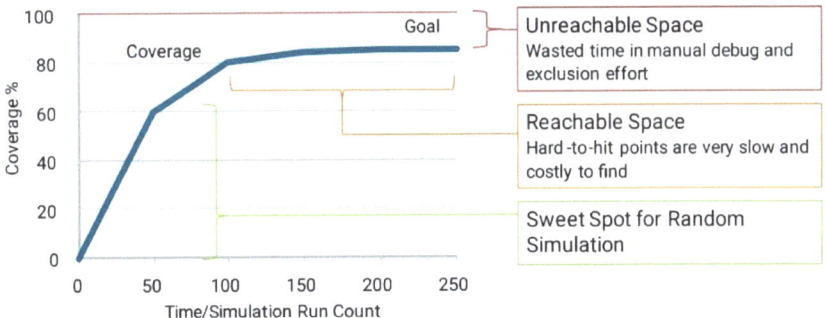

Progress and challenges in coverage closure

An important part of this problem is that, usually, some parts of what seems should be covered simply cannot be covered by any test – they are inaccessible or, in formal terms, unreachable. This might seem odd – if RTL code is not used, why not get rid of it? Because it might be inside a piece of logic which is used in some designs, but not in your design. Or it might be legacy code, lingering in the design because no-one is really sure if it can be removed safely. Remember the engineer's maxim: if it ain't broke, don't fix it, because it's quite likely you'll break something else if you try! So you leave it in there, but you can't ever get coverage on that code in your design and you don't really know what is truly unreachable and what should be reachable if you (eventually) find the right test. Figuring this out is part of what takes so long to get to coverage closure.

This is where a formal app can help – the Formal Coverage Analyzer (FCA) in the case of VC Formal. Formal coverage analysis works together with your simulation environment (with or without existing coverage results) to find logic that is provably unreachable. You exclude this logic from subsequent simulation coverage runs, giving your simulation team a better sense of how much testing they really have left to do. This app alone can have a huge impact on conventional verification effort and schedules.

Formal coverage analysis requires little to no additional input beyond the source RTL. If you already have simulation data, you can use it to focus attention on what has not yet been reached. And you can make your analysis sensitive to all the standard coverage metrics when building unreachability lists: line, condition, toggle and FSM. The VC Formal App can be even turned on during simulation through VCS. Which means most verification engineers are not even aware that formal analysis is being run; unreachable states are simply removed from coverage analysis. How cool is that?

Having the right connections

In modern SoCs, most of the clever functionality has moved to (reusable) IPs/blocks – CPUs, GPUs, peripherals, sensor management, DSPs and others. At the top integration level of the chip, design is now almost completely about connecting all of these functions, and this creates a new verification challenge. We mentioned earlier that reuse emphasizes extensive verification at the IP/block level so that functionality doesn't need to be re-verified at the SoC level. But then how do you verify all the top-level connectivity is correct?

UART	I²C	GPIO	UART2	USB Host 2.0	USB PHY
Flash interface		MPEG		USB OTG 2.0	USB PHY
				S/PDIF	
10/100 ethernet	Graphics processor		CPU cluster 1	SATA 1/2	
DDR2/3			CPU cluster 2	SD/MMC 4-ch	
PCIe 1.1	Embedded memories	JPEG Codec	Audio subsystem	Audio Codec	
Camera interface MIPI CSI-2	MIPI DSI	Graphics processor HDMI TX	General purpose ADC	Touch screen controller	

A block diagram of a typical complex SoC
Each block may have thousands of connections at this level

There are massive levels of connectivity in SoC integration (tens to hundreds of thousands of connections) and, despite that goal of moving all the cleverness into IPs, still quite a lot of logic must be generated to fully manage integration. This includes bus management logic, power management logic, debug, test and IO control logic, and often security and safety management logic. Much of this is created quite late in the design schedule, simply because some of these features can't be completely finalized until the rest of the design is finalized. That's not good news for simulation-based checking which generally requires significant effort to build testbenches; if that effort can't start until the rest of the design is complete, schedules stretch further out.

Certain specialized simulation testbenches can be used to check connectivity (and often are for controls like clock and reset) but these approaches become increasingly difficult to manage as higher levels of configuration control are added. In structures like IO connectivity with many complex configurations, getting to satisfactory coverage can become even more challenging.

A simpler approach is to specify and check this top-level connectivity through a connectivity check (CC) app. For point-to-point connections checks (is this output pin connected to these input pins?) – each connection can be represented as a single tool command. Lot of connections at the top-level are like this and can be checked using these simple commands – often by extracting a list of connections from an early revision of the top-level RTL, manually checking that list then using it to regress later revisions[58].

Some of connectivity required at SoC top-levels is more complex than point-to-point connections; one example is the input/output (IO) logic. Most SoCs have many more internal IOs than can be supported by the limited number of pins on a practical chip package. However, all functions in the SoC typically don't have to be active at the same time so can be managed by multiplexing signals between different function blocks and the IO pads. Architects or application engineers build a complex spreadsheet defining which signals should be accessible at the same time and under what conditions so that all internal functions can be accessed under appropriate settings. From this spreadsheet, scripts automatically synthesize the muxing logic which will manage connectivity between the core and the IO pads.

---

[58] A popular intermediate format for handling large numbers of connections is a file of comma-separated values (CSV), which can be viewed, checked and corrected in a spreadsheet tool.

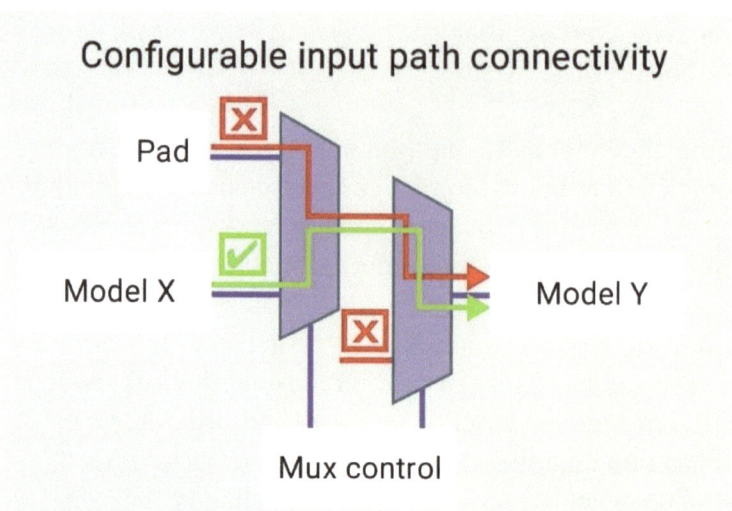

Input path I/O logic - The mux controls determine whether the pad or another internal signal drives Model Y

Checking this logic requires more than point-to-point verification, but these structures are still relatively simple, which makes them very easy to check in formal verification. IO muxing is completely combinational (if you don't include the registers controlling mux selects) and is typically not very deep so you don't have to worry about hitting verification bounds. Assertion generation can be scripted easily from the same spreadsheet, or through a different approach if you want complete independence between generation and verification. And a formal proof will check every possible variation of signal accesses between pads and core under all possible mux control settings, something which would require considerable effort in simulation setup.

While in the IO example logic is (predominantly) combinational, this isn't always the case for logic generated at the top-level. Reset trees in an SoC can be quite complex and may include registers. Debug bus muxing may also include registers, either to hold snapshot values or to help in sequencing wide-debug snapshots onto a narrower debug bus. Even so, these structures still remain relatively simple and are well-suited to formal proving.

Setup for connectivity checking is a bit more involved than for AEP, but still requires no formal expertise. You must create scripts or spreadsheets to define the connection checks, one-line checks for point-to-point connectivity and somewhat more complex specifications for structures like IO muxing or debug busses. In almost all cases, verification teams find ways to generate these scripts from existing files – from a top-level RTL or spreadsheet for point-to-point connections, or from specification spreadsheets for structures like IO muxing. Naturally this implies need to verify the scripting and the first-pass spreadsheets are correct, but many teams still claim high value and total engineering effort saved in this checking, especially in regression, even for SoCs as small as 20M gates.

Sequential equivalence – some assembly required

Our next example takes us one step closer to hands-on formal while still limiting the complexity of analysis. In this case the app will create the required assertions for you (given some simple inputs), it will do the run and it will present you with results in the standard debugger, but you start to need to get a little more involved in the proving process, in part by adding mapping points which can help the prover identify points of correspondence between the two versions of the design. You may also need to get a bit more hands-on if proofs in some cases do not complete (are inconclusive).

First, what is sequential equivalence checking, why is it important and why isn't it covered in standard equivalence checking? We talked about equivalence checking in chapter 2. That type of check is used to verify that the gate-level model generated though synthesis is functionally equivalent to the source RTL (that is, that synthesis didn't break the functionality). Because of the way synthesis works, this only needs to check equivalence of combinational logic between registers[59].

However, some design changes require sequential modifications. One such case involves adding clock gating for power optimization. The ultimate functionality should remain the same in the enabled state, but cycle shifts (latency) may be added when enabling and disabling the clock since new register stages have been introduced. What must be compared between the "before" and "after" logic now includes logic which would confuse conventional equivalence checking[60].

---

[59] This isn't strictly true. Synthesis engines may re-time logic across registers, so equivalence checkers have to understand these cases also. But allowed deviation from the basic principle is limited.
[60] The gated and ungated functionalities are logically different but in a well-controlled way

Clock Gating Changes for Power Optimization

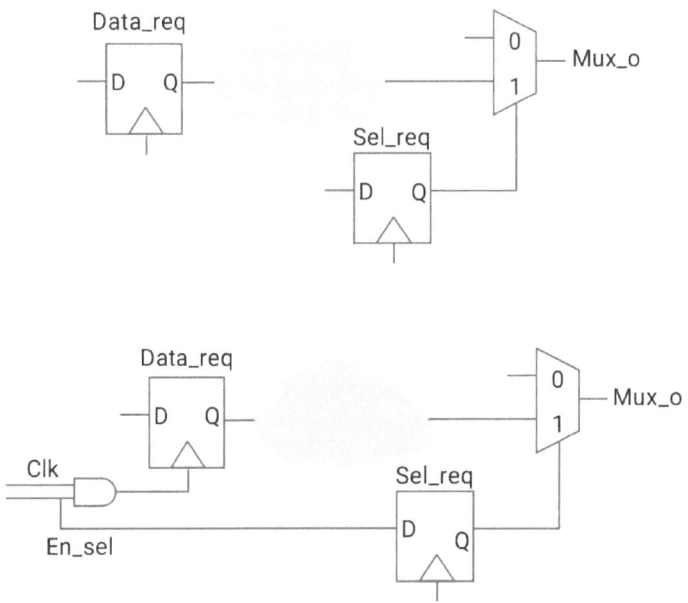

Clock gating optimization – before gating (above) and after (below)

Since sequential equivalence checking is about comparing two designs, you need a source RTL (before you make changes) and an implementation RTL (with clock gating changes). Just as when you compare two versions of the same Word document, you'll generally see these two RTLs side by side in the debugger window. The assertion generation part of SEQ is still hands-free – the app will generate these for you.

The debug part of SEQ (looking at results) is based on Verdi and is going to look very familiar to anyone who uses that tool for simulation debug. You look at detected problems which leads you to waveforms, then you'll cross-probe to RTL and trace back to root-causes. The only aspects that are a little different here are that you are looking at two sets of data (the specification/original version and the implemented version) and there are some extra debug features in Verdi, such as sequential traceback to a root cause. But all of this will still be very familiar to a verification engineer.

The only part of the task that gets more (formally) hands-on is the proving phase; you can think of this as a first introduction to the details of formal proving. In conventional equivalence checking, comparisons are quite bounded – tracing most typically stops at sequential elements like registers. But in sequential equivalence problems, as in clock-gating verification, tracing may have to go through multiple sequential elements. This makes the potential complexity of the problem very large, which in turn can result in proofs which do not complete in reasonable time or memory. Proofs terminations of this type are our earlier-mentioned *inconclusives* – no counter-examples have been found as far as the proof was able to reach, but equivalence has not been conclusively proven.

SEQ provides a nicely automated way to overcome this by automatically decomposing challenging problems into smaller sub-problems, then proving equivalence between these sub-problems. You can track this progress in Verdi. Where sub-problems converge, those sub-proofs are complete. Where they fail to converge or where an apparent mismatch (assertion fail) is found, that signals need for a new (and automatically triggered) decomposition. The app will continue to try to find new decompositions, so this part continues to be largely hands-free (you can still watch progress/status in Verdi). In the simpler cases, the sub-problems converge, leading to a proof (or counter-example) at the full problem level.

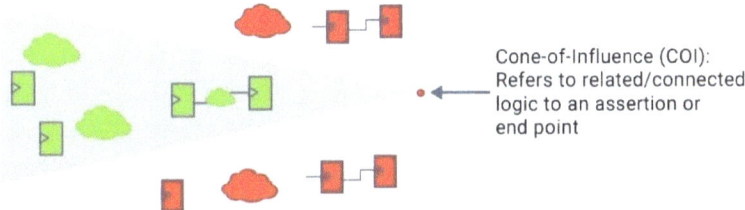

The cone of influence (COI) of a property
Light-colored logic is in the cone of influence, darker logic is not

Where you may have to get involved is if it becomes clear that certain parts of the design are resisting convergence. This could be caused by memories or counters or other complex blocks in the **cone of influence** for a proof, since these typically lead to explosion in size in formal methods. At this point you may need to turn to the tools we mentioned earlier (abstraction, invariants, constraints). If you are planning to be a hands-on verification engineer, fear not; learning how to use these techniques doesn't require an advanced degree; this is just more skills development to add to your arsenal of verification expertise

Other apps
The apps discussed above are representative of a range of usage, from easy to a little more involved. But they aren't the only apps you are likely to use. We won't go into as much detail on other apps, but we will give you a flavor of what is commonly available:

- **Register verification** – almost all CPU-based designs today are memory-mapped. Components in the design are controlled, written to and read through registers and that register logic can be quite complex – registers which are read/write or read-only or are cleared on read or many other possibilities. They come in different sizes, and different ranges (bitfields) in those registers have different functions, each with that wide variety of possible read/write properties. And there are more variations. Getting this right isn't just important to hardware verification. Software communicates with the hardware through these registers, so absolute correctness in behavior is vital. Register verification apps automate this checking against an XML or similar specification.

- **Design exploration or navigation** apps are useful in several contexts:
    - For an RTL designer in exploring aspects of the behavior of a design without needing to create a testbench or to have a detailed understanding of property checking
    - For a verification engineer in building assertions and constraints, to explore what properties flag as violations or as valid
    - In post-silicon debug to trace-back behavior found in the lab to likely root-causes.

- **X-propagation analysis** apps: In simulation, 'X' is used to indicate an unknown state (rather than 0 or 1). Registers which are not reset on design initialization will start in this unknown state. In some cases, this is not a problem; the state is set to a known value before the value is needed. In other cases, if this state is not reset an unknown value can cascade through subsequent logic causing serious misbehavior. X-propagation analysis looks for and flags all potentially serious problems.

- There are other apps, around security analysis and signoff, which we'll talk about later in this book.

Are we there yet?

These apps provide valuable verification needing only limited learning, but they don't represent all that can be accomplished with formal verification. Use for more complex or more unique problems requires you to write your own assertions and may require some level of involvement in getting to proof-convergence through abstractions, constraints and other methods. We'll dip into this topic later in the book.

Also, while we've said this before, we'll repeat it again. A characteristic of the bounded model checking at the core of most formal methods is that while a proof (an assertion passes or fails) or a counter-example will often be found within the bound of proving, this is not guaranteed. We started to see this in our short introduction to the SEQ app. In these cases, to get to a proof, you can increase the bound, or do more abstraction, or add more constraints, or any combination of these. Or you can ask whether the depth to which you have checked, with the abstractions and constraints you already have, is sufficient to declare the proof acceptable. We'll look a little more at this topic later in the book.

Now we're caught up with the basics, it's time to look at how your peers in the semiconductor industry are using formal verification today.

The role of formal in design today

Adoption

No matter how clever or easy to use a technology may be, the only measure of success that matters to both vendors and users is real-world effectiveness, as indicated by adoption and growth; how many people / organizations are using it and how quickly is that usage spreading? Multiple surveys in the industry indicate that both are robust and have moved beyond early expert adoption.

For example, one survey[61] shows about 20% of reporting projects using formal apps in verification and around 35% using custom property-checking. Among these, app-based verification contributed to significant growth in formal usage from 2012 onward, followed by notable pickup in custom checks from 2014 onward, indicating perhaps that growing familiarity with app-based approaches is making verification teams more comfortable in moving also to those custom applications.

You might be surprised to hear that formal verification is picking up even in FPGA-based design[62], at around 10-15% of projects in 2014, maybe reflecting the increasingly complex nature of FPGA SoCs and the need for development teams to check *cannot-fail* assertions as completely as in ASIC designs (especially in mil-aero applications where FPGAs are widely used, and now even

---

[61] Verification Trends 2016
[62] 2014 Wilson Research Group Functional Verification Study

in ADAS, where safety and reliability expectations are also very high).

Regarding business growth, informal feedback supports a view that the top 20 companies (by revenue) designing complex SoCs are already investing $50M/year in their formal verification flows. Perhaps twice that much is being invested across the entire user-base. Also noteworthy is the range of applications; designers are applying formal methods across the spectrum, from processors and graphics, to wireless, networking and storage, image processing and recognition. Indeed, there is no obvious reason not to use formal in most areas of (reasonably complex) digital design today. That said, designs under ~100k gates (e.g. sensor logic) are not reporting formal usage, which isn't really a big surprise.

Among organizations that have established teams dedicated to formal-proving, it is common to hear that these methods now address 20-30% of the total verification burden[63]. Formal is no longer a niche technology – it is now carrying a verification load similar to other verification methods.

An important question is **why** these teams are using formal methods. A cynic might hold that "it's hot so we better do some to keep management off our backs". We doubt this accounts for any significant usage; most verification teams are under such intense pressure they don't have time for "show" projects. A much more

---

[63] See the foreword, also Simulation and Formal – Finding the Right Balance

common reason heard from verification leads is building concern that as design complexity continues to grow they already see or anticipate important verification problems moving beyond the reach of simulation-based methods. They are allocating budget, resource and schedule to ramping up in formal because they have no choice. Rare or intermittent problems which escape to the field are often out of reach of "just try harder in simulation" but can spiral into disproportionate costs and reputational damage. Design and verification teams are increasingly turning to formal methods[64][65][66][67] as one way to shake out those difficult problems.

Size constraints

Whenever formal methods are discussed, you will hear some dismiss them as "only for small problems". The exhaustive nature of formal proving certainly limits the size of the state-space that can be considered in a proof, but this should be offset against the potential for abstraction, automated in many high-value apps, and of course continuing improvements in proving technologies, especially around advances in heuristics.

Given that, what are realistic size-limits? It depends. A full SoC state-space far exceeds the practical limits of any current formal method, but there are multiple useful SoC-level problems which are routinely tackled, especially in

---

[64] [Adoption, Architecture and Origami](#)
[65] [System-Level Formal](#)
[66] DVCon 2017 Making Formal Property Verification Mainstream: An Intel® Graphics Experience
[67] DVCon 2018 Architectural Formal Verification of System-Level Deadlocks: Qualcomm and Oski

apps like connectivity checking where block abstraction is easily automated. Cache-coherency verification is an example of a system or large-subsystem-level problem also commonly reduced to a manageable level through abstraction; we'll talk more about this later.

Unquestionably custom property-checking is most commonly used at the IP level, since pre-tested constraints and heuristics may not be available and therefore a limited state space makes proving less challenging. Even at this level, abstraction is often needed to simplify memories and datapath elements. However, this need is really a feature of the formal approach rather than a limitation. You do this in order to enable formal proving on other parts of the logic, remembering that simulation may also be limited in exhaustively proving through such cases and often depends on abstractions such as behavioral models.

Granting all these points, what is the real story on state-space capacity in formal tools? This is a moving target and you'll no doubt hear different ways to calculate from other sources, but one approach we like for its simplicity is this. First, size limits are a function of the property being tested. The size of the cone-of-influence (COI) of that property is much more important than the design size.

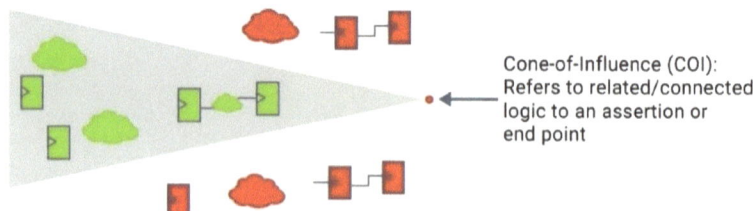

Cone-of-Influence (COI): Refers to related/connected logic to an assertion or end point

Reminder: Cone of influence for a property

Given this, the problem size of the proof you are facing is determined by the number of inputs and state elements in that COI. Formal tools will often calculate these value for you. Within these bounds, one experienced user in the industry[68] cites **40K state elements** as a practical limit as of 2017.

Popular targets for property checking

Clearly, most teams start and continue with apps. These add obvious value in covering important parts of the verification plan (connectivity, coverage, registers, etc.), they're an easier place to start than custom property-checking and they help you incrementally build expertise in using formal tools. Adding custom property checks comes later and is commonly viewed as an incremental extension to app-based checking to handle specialized and proprietary objectives.

Good targets

When looking at block-centric checks, whether app-based or custom, some objectives fit well with formal while others are better handled in simulation. In general, formal works well with control-intensive blocks which have lots of states and transitions, with complex conditions for transitions and possibilities. Think of very complex FSMs or multiple inter-operating FSMs. Common blocks in this class include arbiters, controllers of various types (memory controllers, power management controller, traffic controller, interrupt controller, etc.) and schedulers (round robin, fair etc.) for quality of service.

---

[68] Erik Seligman in Formal Verification: An Essential Toolkit for Modern VLSI Design

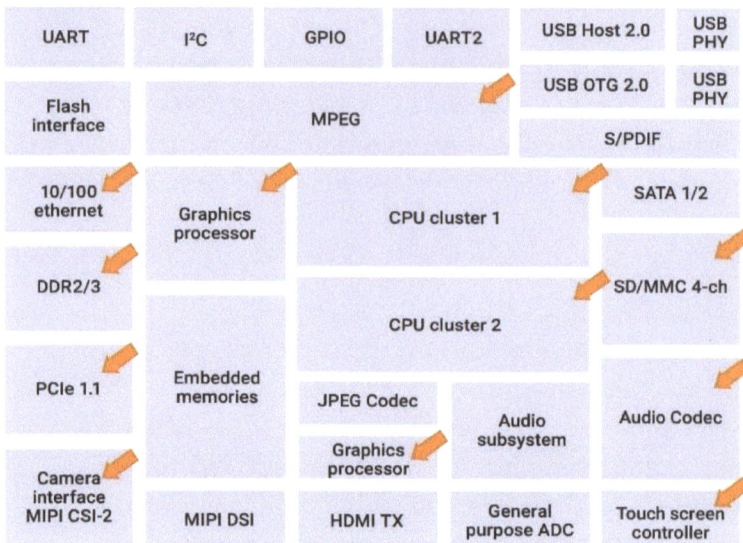

A representative SoC – arrows indicate some areas formal will likely find a role

Similarly, in networks, verification of bus bridge behavior (from master to bus, from slave to bus or from one bus protocol to another) is a good candidate, ensuring that data isn't lost due to FIFO overflow in the bridge, or loss of request or acknowledge signals from/to the bridge. Equally proving correct operation of network traffic management techniques, such as token management, are excellent candidates for formal proving. Cache coherent networks are a very hot topic in modern multi-processor systems and pretty much depend on a significant level of formal validation (we'll have more to say on this topic later in the book).

Data transport systems are good candidates, where data is being moved from one place to another rather than being transformed in computation. Examples can be found in crossbar switches and networks-on-chip (NoCs).

An area of verification that is always of high interest for formal methods is interface checking, especially for protocol validation. This usage should be approached with some caution. Verifying simple interfaces is not too onerous; for example, coding assertions and constraints for the AMBA APB protocol can be completed in a matter of hours. However, checking for more sophisticated interfaces can become very complex very quickly; coding a correct and complete set of assertions and constraints for AXI3 (or higher) could take weeks.

The best approach for interface-verification is to use pre-packaged assertion IP whenever possible; these are already setup with assertions and constraints proven to work effectively in a wide variety of contexts. Supporting documentation will tell you what aspects of operation are covered and what (if any) you need to handle separately in simulation-based testing.

Targets to avoid

Datapath functions (adders, multipliers, filters, etc.) can be challenging for formal methods since they typically expand into gigantic state graphs, particularly thanks to the need to do bit-level analysis on word-level values. That said, this area continues to see advances in proving methods. Specialized provers are already known and even available in some cases – watch for these techniques appearing in commercial tools[69], also see the "Looking forward" section in this book.

---

[69] We touch later on induction methods as one way to address this class of designs.

Similarly, complex data transformation functions like JPEG, MPEG and encryption / decryption are simply beyond the capacity and performance of today's formal engines, at least at the fully elaborated design level. However, intelligently abstracting the quite regular architectures of these designs could put aspects of their behavior within the reach of formal. This is another area to watch for improvements.

Certain tests on interfaces which require proving over very long sequences (such as for PCIe, MIPI, SATA or HDMI) are not formal-friendly for an obvious reason. Standard formal methods depend on proving/disproving within relatively limited clock-cycle bounds; if a proof depends on analysis over sequences of many thousands of clock-cycles, the state-space can become unmanageable. Even here there are already bug-hunting methodologies for directed searching across many cycles.

One more example. Formal methods are naturally designed to provide binary (pass/fail) responses rather than statistical responses. They are not well-suited to quality-of-service (QoS) verification, except at the fringes. If you want to know whether your system can ever deadlock, or will acknowledge a request within some time bound, formal can help. But if you want to know whether each requestor is handled fairly across different traffic profiles, that is a problem for simulation and statistical analysis, not a problem for formal.

Product and service solution providers
Apologies to any solution providers we may have missed; we believe this list is fairly exhaustive.

Synopsys (VC Formal)

Synopsys have built their formal product lines entirely in-house, originally represented by their Magellan product, later adding the Hector high-level equivalence-checking tool. They subsequently developed VC Formal as a completely new product, which now has largely superseded Magellan. Synopsys has a broad product and solutions offering.

Cadence (Jasper Gold)

Cadence started in this space with BlackTie from their Verplex acquisition, evolving into Incisive Formal Verifier. Later they acquired Jasper Design Automation. Jasper is generally credited with raising the profile of formal verification, in part through intensive customer support/services model and in part by popularizing the app concept. Cadence has a broad product and solutions offering.

Mentor (Questa Formal)

Mentor became prominent in formal through their acquisition of 0-In, which made them the leading player in formal in the early days. They added further formal capability in sequential equivalence checking and C to RTL equivalence checking through initial engagement with Calypto and later acquisition. Over time they have rebranded this solution as Questa Formal. Mentor has a broad product and solutions offering.

OneSpin (360 MV family)
OneSpin was (appropriately) spun out of Infineon from an internal formal verification team. OneSpin has a broad product and solutions offering.

Oski (services and verification IP)
Oski is primarily a formal verification services company. They provide methodology setup services, training, joint development, verification kits and IP.

Others
Real Intent started with a strong focus in formal verification. Market presence is now primarily around their static verification product, while formal capabilities are centered in the Ascent family.

Averant provides formal verification software for automatic design checks, sequential equivalency checking, property-checking and timing constraints checks

Atrenta (prior to acquisition by Synopsys) used formal engines in SpyGlass in application-specific uses, for example in CDC analysis and in timing constraints checks.

Adding formal to your flow

This section is written primarily for verification managers and verification team leaders, to guide managing formal verification in your organization, **without having to be a formal expert**. We're not going to spend any time on how to setup and run tools; we are going to spend a little time on how to plan for formal in the larger testplan, how to understand and guide formal team members and how to measure progress towards a signoff goal. That said, even though this isn't about tools, methodologies may vary between vendors. Again, we'll be guided by the Synopsys *VC Formal* flow. Many of these concepts will carry over in some manner to other flows but you should check with your vendor for possible differences.

Organic skills growth

Despite the obvious advantages in adding formal methodologies to a verification flow, simply jumping in as you might with any other new capability (buy the tool, train on the tool, start to apply on live designs) doesn't always work out very well, in part because formal requires a mental switch from familiar simulation methods, also because it may be viewed with suspicion by the rest of the verification team. On the other hand, postponing adoption until you are able to hire formal experts (who we'll call black-belts) generally becomes a decision not to make a decision, and that can be an expensive mistake if you already anticipate hard problems overwhelming simulation-based flows.

Formal black-belts are rare today and have little difficulty finding jobs with big-name companies. If you want to build formal capability in your team, you almost certainly need to grow it internally and/or recruit whatever brown-belts you can find. Experiences learned from companies who have done this successfully suggest some common steps towards adoption:

- **Encouragement**: Verification team members are encouraged to learn more about formal verification successes, applications and case studies through conferences, EDA vendor meetings and personal learning. Make a little white-space in those crushing product schedules.
- **Find and grow a champion**: When one of those people shows an interest in championing formal methods, encourage and support them with a small team, budget, schedule and supervision for planning, discussion and feedback. This shouldn't be difficult - most engineers want to learn new

skills. Most important, even more than background in formal, is willingness to take risks.
- **Freedom to learn and develop a methodology**: Managers should provide plenty of flexibility for learning and finding a path to demonstrate initial successes. No-one gets hung up on "we already showed that in simulation" or "we got there much faster using simulation". The goal is to learn and develop a working flow, especially for the apps. Experienced verification teams have been quite open about this phase taking a year.
- **Pick the best targets**: We'll say it again – start with the Apps and AIPs. These are by far the easiest path to develop expertise and show value. Over time you can graduate to adding custom property-checking, but there's no need to rush to that point.
- **Develop metrics**: It is always important to develop metrics for what is being proven, but early flexibility must be allowed in these metrics – teams need to grow into what they can prove and should be allowed to start with simple metrics like "no unresolved counter-examples". Capture data on effort (engineering, CPU hours, etc.) to help develop ROI cases. Over time, push for harder limits, e.g. around coverage.
- **Hands-on big picture supervision**: An important point here is that the verification supervisor (you perhaps) should not be completely hands-off. You don't necessarily have to run the tools or understand the details, but you do need to keep connecting metrics and success back to the big picture – is this ultimately headed towards production usefulness? At the same time, you can become more familiar with the concepts of the

domain, so you can learn (at a high-level) how to question approaches and provide guidance.

- **Socializing in the design and verification teams**: It is also very important to socialize progress within the larger verification team and to gather constructive feedback. Fostering a collaborative environment with the formal team ensures everyone understands goals and benefits and helps the formal folks optimize their focus. You definitely don't want mainstream verifiers looking at this as low-value playtime. There is also real benefit in including designers in these discussions, first because formal verifiers may need their help in handling inconclusive proofs. If designers are not bought in, they will not provide adequate guidance to get to formal signoff. Second, there is value in promoting use of formal in RTL design; the formal team can help designers use the tools to explore behaviors in their RTL, ultimately helping them to hand-off higher quality IP to verification.
- **Socializing in the management chain**: Successful adoption efforts have also been careful to socialize progress and goals further up the management chain. You might be surprised – even CEOs can be interested in what you are doing in formal, especially where it helps add to their product quality, safety or security pitches.
- **Review, refine, advance**: As in any good engineering project management, the first round should be followed by careful review and analysis: what worked well, what didn't, what should be attempted next and how metrics should be tightened to more effective levels. This review should definitely involve the larger verification

team; the expectation should be that in the next round the formal team will contribute materially to verification closure. That said, be patient. Getting to productivity may take a few false starts. Once you start getting successes, that success will be contagious.

First-cut targets for formal verification

The essential components of your verification process don't change when you add formal methods to the mix. You build a verification plan, partitioned by blocks and behaviors to be verified. Within that plan, you will segregate the verification/coverage plan by appropriate verification technologies:

- Goals that should be a good fit for formal
- Goals that will work well with constrained-random (IPs, small subsystems)
- Goals that will work well with emulation (regressions, SW/HW co-verification)
- Goals that will work well with FPGA prototyping (regressions, SW/HW co-verification)
- Goals that must use simulation for other reasons (eg AMS verification)

When looking at candidates for formal, consider these cases:

- Any verification task for which an app already exists, like top-level connectivity checking, register checking and sequential equivalence checks around clock gating. This is the easiest place to start; it should be much easier than simulation and it will be more complete. These cases should be no-brainers.

- Verification plan line-items for which assertion IP are already available, for example the section(s) in the verification planning covering protocol compliance on interfaces such as the AMBA interfaces (ACE, AXI, AHB, etc.) and common IO standards (USB, aspects of PCIe, etc.)
- Any case where you know control complexity is very high and difficult to cover solely in dynamic testing (interdependent FSMs are generally good candidates)
- Blocks which have historically been problem-prone from release to release, exhibiting intermittent hangs, deadlocks or other issues

And of course, remember good targets are often more about the test than the block. Any given block may be best served by leveraging multiple verification platforms, each targeting different tests. The "challenging blocks" case is a good example. Maybe these could benefit significantly from some carefully-crafted formal proving in addition to dynamic testing.

Detailed planning

Once targets are identified, then you start planning detailed tests. While there is a concept of testbenches in formal verification, this term is primarily associated with the end-to-end verification concept we mentioned earlier in the book. Since this is a fairly advanced usage, you probably won't consider it in early stages of formal adoption. In app-based verification and your initial forays into custom property verification, this testbench concept is not so obviously relevant so we'll stick to calling these tests.

Just as in detailed planning for simulation, you should have a planning stage per block (viewing top also as a block). In each case, you want to start with plain English descriptions of checks you want to perform, for example "verify this state machine always returns to IDLE after interrupt". Get these descriptions right first, because they'll be easier to understand and debate than the SVA translations they will eventually become.

Once detailed planning for a test is complete, your formal experts can start building an executable test. This won't look much like the UVM bundles you know (and maybe love), running to thousands of lines of code. You should expect a few Tcl files, SVA files and bind files, in total running to a few hundreds of lines (lines of code similar to targeted simulation setups). Again, you can get to testbench concepts in formal verification, but that is definitely for more advanced users.

Monitoring progress

In simulation you ask team members to provide regular updates on how many bugs they found and what coverage they have reached. Metrics aren't so very different in formal verification. The team will find counter-examples (CEXs/bugs), some in the design and some in their tests. They'll find these quickly at first, then more slowly as bugs in the design and in tests are shaken out – this should sound very familiar.

A nice surprise for verification managers is that formal verification will start finding bugs very quickly; you don't have the typical simulation ramp-up phase of getting the testbench working. You may also discover that formal will uncover bugs in lots of unexpected places. While simulation testing works deliberately through a plan, formal races out to the fringes and can expose bugs before simulation test; a very real plus for accelerating verification closure.

In other cases, the formal team will report completed proofs for some properties and inconclusive results for others. This is where progress monitoring diverges from simulation, getting into questions of whether a completed proof is valid (related to constraints that have been set for that proof) and what steps can be (or need to be) taken next with inconclusive proofs.

The metrics you want to monitor, by verification plan line item, are:
- How many properties have been developed (asserts, assumes, covers)
- How many failures and covers have been found, and trends on these

- How many completed proofs and trend (but first read the next section)
- How many inconclusive proofs and trend (but first read the section on bounded proofs)

Tracking progress on developed assertion, constraint and cover properties

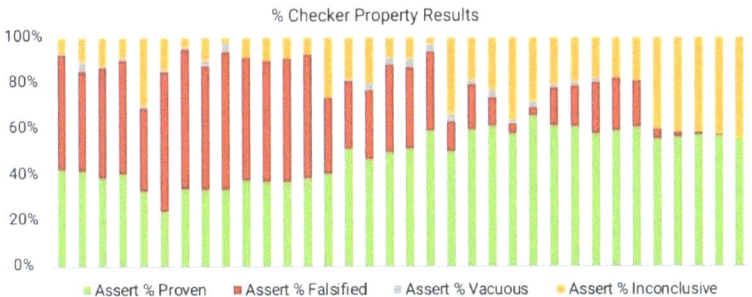

Tracking progress on assertion proof-status

Under-constraining and over-constraining

There's a Goldilocks aspect to constraints. You don't want too few, you don't want too many, you want just the right constraints. Under-constraining can cause formal engines to explore parts of the state space which are not meaningful for any practical use of the DUT (or perhaps for the usage you intend). That can have bad consequences. Analysis may report spurious counter-examples (CEXs), which are failures in ways that are unrealistic. Or analysis may simply fail, running out of time or memory. Either way this takes engineer and machine time to analyze and to debug while not really advancing coverage.

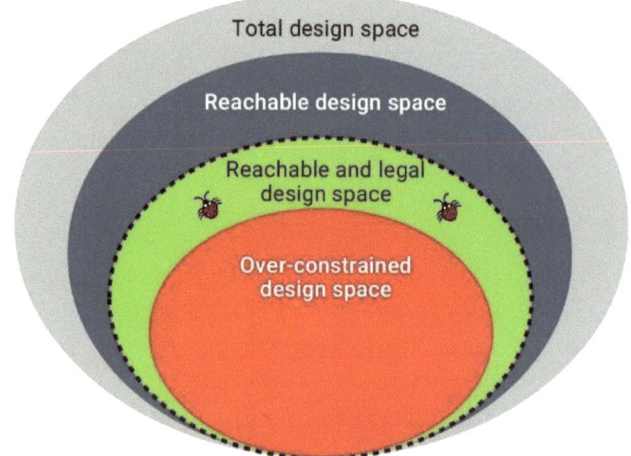

*The smallest area is an over-constrained
state space, missing some bugs
The next larger area is the ideal –
a reachable and legal space*

Good understanding of the modes of operation of the DUT or protocols or handshaking used by the blocks is the best way to avoid this problem. If you're finding bogus CEXs or a run won't converge, perhaps you need to add constraints – after discussion with the designer or architect. Where appropriate, one suggestion is to use pre-built Assertion IP (AIP) which have built-in constraints. This will help avoid spurious failures and get you up and running more quickly.

On the other hand, you can over-constrain, creating conditions which prevent exercise of realistic behaviors. There are some legitimate reasons to over-constrain during the course of setup or bug-hunting, but not in signoff. In the VC Formal flow, you can check for this possibility by running over-constraint analysis for each property. You can also check using reachability analysis for cover properties you know you should be able to reach; if you can't reach them, an over-constraint may be to blame.

Bounded proofs (inconclusives)

As we have (no doubt tediously) repeated again and again, inconclusive results are a fact of life in formal verification. Remember that verification is hard; there are no silver bullet methods to automatically verify everything. Formal methods provide a way to exhaustively verify properties out to some defined clock-cycle depth (proof-depth). Which means that it is possible for such a checker to hit that bound without finding a counter-example and without completing a proof that the check passes.

In earlier times, inconclusives were considered a real barrier to formal adoption – if a result was inconclusive, surely that meant it wasn't useful? Over time a more constructive and actionable viewpoint emerged[70], starting with a view that these were **bounded proofs** rather than

---

[70] See for example "Signoff with Bounded Formal verification proofs", best paper award in DVCon 2014: http://events.dvcon.org/events/proceedings.aspx?id=163--2

inconclusives. It's now much more **common to view a bounded proof as fully acceptable for signoff, as long as the bound is well justified**[71]

This is understandable. The tool has exhaustively proved the absence of a counter-example out to the proof-depth; that is already valuable information. It is very likely that at some increased proof-depth, the proof would be conclusive. In some cases, it is quite possible that this proof at this depth is already sufficient or would certainly be sufficient if extended out to a slightly deeper search.

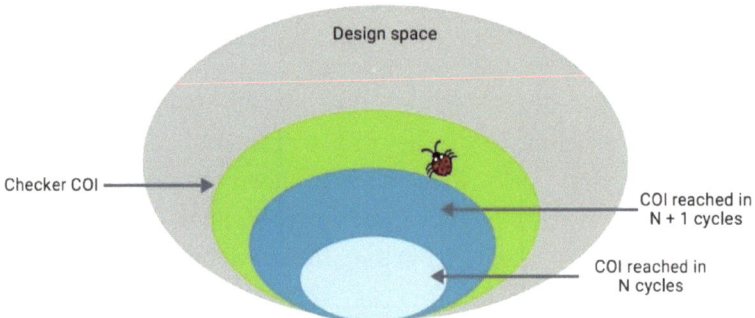

Exploring limits on a bounded proof

Of course, you can't just hope this is true; you need to know if the proof you have is practically sufficient or if you have to work harder. A starting point is to look at a bounded coverage analysis, ideally also providing an incremental cover analysis. Coverage as a percentage of the property cone of influence is a good starting indicator. If this isn't high, look at the delta in coverage over the last

---

[71] Formal Verification: An Essential Toolkit for Modern VLSI Design

two steps. A significant increase suggests that it may be worth trying again with an increased proof depth.

Another commonly-used technique is to create cover properties for "interesting" corner cases. Finding what proof-depth is required to hit these properties can be a good indicator of a minimum required depth[72].

Whenever you are exploring what depth may work, eventually you should have a discussion with the designer to agree on the "design/property radius" (the depth) they would consider acceptable. If the bound is not practical for formal proving, then you have to consider abstractions to reduce the complexity of the problem space; that's the subject of the next section.

Manually-guided proofs

It is quite possible that in some cases, an attempt to prove a property on a design will fail to complete because a full proof would exceed time or memory limits. We talked earlier about various methods to manage these cases – abstractions, decomposing a problem with assume-guarantee properties at interfaces between the sub-problems, and case-splitting with constraints.

You should expect that your team will need to resort to this kind of guidance in more than a few cases. But you'll

---

[72] If you want to test your skill, you can try estimating the maximum number of cycles required to prove an assertion - with knowledgeable design help, of course. You may find in this exercise that you are missing an important constraint or two – which might further help in bringing proofs to closure.

be happy to hear that formal tools now go a long way towards simplifying this task. For example, tools can often auto-detect candidates for abstraction, such as datapath elements (though replacing logic with an inferred abstraction typically requires engineer approval). Decomposition and managing the assume-guarantee flow is also commonly supported by extensive automation in apps, a major plus in managing problem size challenges. Still, even with these advances, closing on some proofs will continue to require hands-on effort and discussion with design and other verification teams.

Getting to formal signoff[73]

Ultimately you and your formal team should assume responsibility for signing-off meaningful components of the verification plan. To do that, you first need to measure progress against a set of goals. We suggest the following table as a starting point, though you can certainly adapt (and evolve) this to best suit your needs. These metrics range from simple but relatively low confidence to more complex with increasing degrees of confidence towards signoff quality. Formal tools should help your team gather these statistics. And of course, you will want to trend these statistics during the evolution of projects.

---

[73] Here we are not talking about formally signing off a complete block or IP. That is another interesting topic, but beyond the scope of this book.

Metric	Degree of confidence provided	Current project status	
		Example target	Actual
Property density	Low	1 assertion / 20 lines of code	
COI/Lines covered by assertions	Medium	90%	
Logic/lines covered by formal core	High	85%	
% (mutated) faults covered by assertions	Highest	95%	

Metrics to use in assessing progress to signoff

In common with simulation, formal tools typically support metrics based on (RTL) line, condition/branch, signal toggle and FSM state coverage. You can select any one of these metrics (in the table we use line-based metrics as an example) to give you a measure of completeness of coverage when performing analysis. The value of these metrics is to provide a much more detailed and functional assessment of assertion coverage than you got in early assertion-density analysis and is a much more concrete measure of verification progress than the simpler assertion-density metrics.

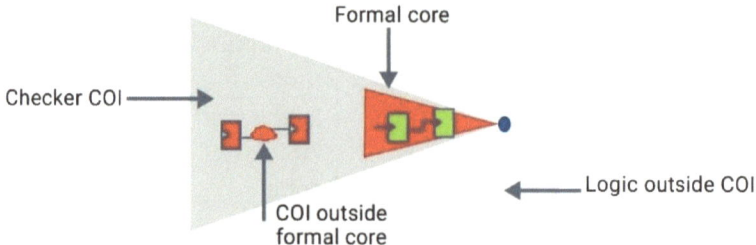

A formal core is usually smaller than the cone of influence

A more precise measure of formal coverage is based on **formal cores** (also known as proof cores). A formal core is the portion of logic **required** to prove a property and is generally a subset of the cone of influence for that property.

All of these coverage metrics provide a way to cross-check between what you (or the designer) believe should be touched in an adequate proof versus what you find the tool tells you was really touched. If these correspond, your confidence goes up. If not, perhaps the test needs to be further refined with more assertions or cover points, or perhaps some constraints need to be relaxed.

Another very useful cross-check applies RTL mutations to insert bugs into the DUT (this is provided in the VC Formal FTA app). You would expect that, under a mutation, at least some assertions should fail. If they don't, this suggests that assertion coverage should be improved, or perhaps that proofs are over-constrained. By running through a number of mutation runs and correcting any exposed coverage problems, signoff quality should become even more secure, delivering the confidence you ultimately want in those proofs.

How you ultimately choose to define signoff based on these metrics should not, conceptually, be very different from how you do this for dynamic signoff. We're just more accustomed to the process for dynamic verification. First you need metrics – we've already discussed these. You want to know that all the assertions you are able to prove have been proven with no open counter-examples, and that cases which resist formal proving are passed back to the dynamic team for a different angle of attack.

You want to know that for each of the proven cases you have good coverage, especially for the strictest levels of checking (formal core) and that no proofs were over-constrained. You should have similar expectations for bounded proofs, with the added requirement that each bound is validated by the designer and/or reachability tests for appropriate cover points. And finally, you will trend metrics to determine where no further progress is being made in formal checking – that you have done *enough* within the bounds of the tasks you assigned to formal.

Again, not so very different from how we approach signoff in dynamic verification. Of course, you will want to build experience and confidence, in order to decide where you want to set the bar. When you get there, formal signoff should be just as certain as dynamic signoff.

Looking forward

Now you understand the basics of formal verification, where can this technology take you next? There are many ways of looking at what might come next in any technology, from logical next steps to deeply technical advances and even wildly-speculative potential. We decided to focus on areas that are probably more immediately useful to our audience, starting with applications that go beyond those discussed earlier to address very topical concerns for current SoC designs, and then to look at near-term technical advances aiming to further enhance the usability and reach of formal verification.

Application domains

Many of the following applications are already available in some form, though they continue to advance in capabilities. Each builds on an application-aware understanding of a problem to construct appropriate assertions, limit the scope of checking to increase the likelihood of closure and provide application-aware debug. We'll start with a couple of widely-used applications then get into some emerging or more advanced use-models.

Functional clock domain crossing analysis

SoC design and CDC analysis go hand-in-hand. Any device supporting (at minimum) multiple peripheral interfaces must support multiple clock domains, implying lots of clock domain crossings and need for care in managing metastable states and lost data at those crossings. Many design or verification teams look only at structural analysis of these crossings; that analysis is very important but can be further enhanced with functionally-aware analysis.

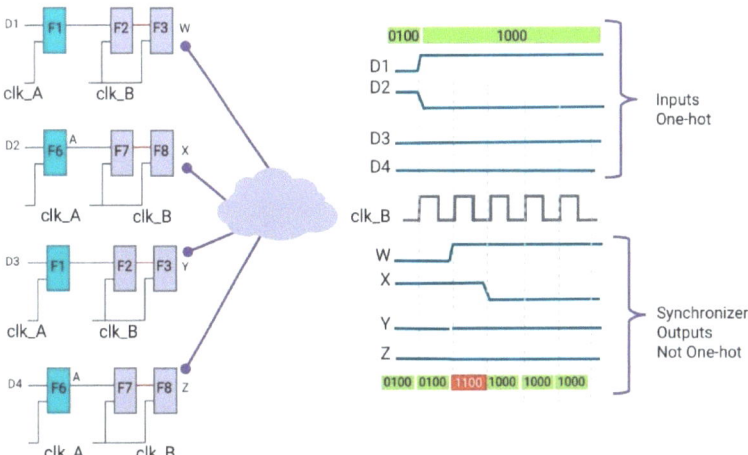

When multiple paths between two clock domains converge they must be one-hot encoded to guarantee correct operation

One example check considers the correctness of data transmission between domains. Conventional synchronization deals with metastability at crossings, ensuring that a register at a crossing doesn't lockup under certain conditions. However, this alone doesn't ensure that transitions may not be dropped or shifted in a crossing (for single bits) or may not become temporarily invalid (for vector signals).

Handling cases like this requires careful design, using handshakes or grey-coding or one-hot coding for example; formal verification is a great way to fully check that this crossing-management logic has been implemented correctly. The more advanced CDC tools will offer these kinds of checks alongside structural checks.

Functional power intent verification

Mobile and green technology demands, costs and reliability have all driven power to become a high-priority in delivering competitive designs. This led to creation of the UPF standard for defining power intent to drive power-aware implementation. We quickly discovered that these intent descriptions for SoC designs can be extremely complex, which in turn prompted new verification methodologies to validate the correctness of the intent.

Some of this intent can be verified statically. When crossing from one voltage domain to another, power intent should require a level-shifter. Or an input to one block from another block in a power-switchable domain should have power intent specifying isolation logic. These needs are not use-case dependent so can be checked statically.

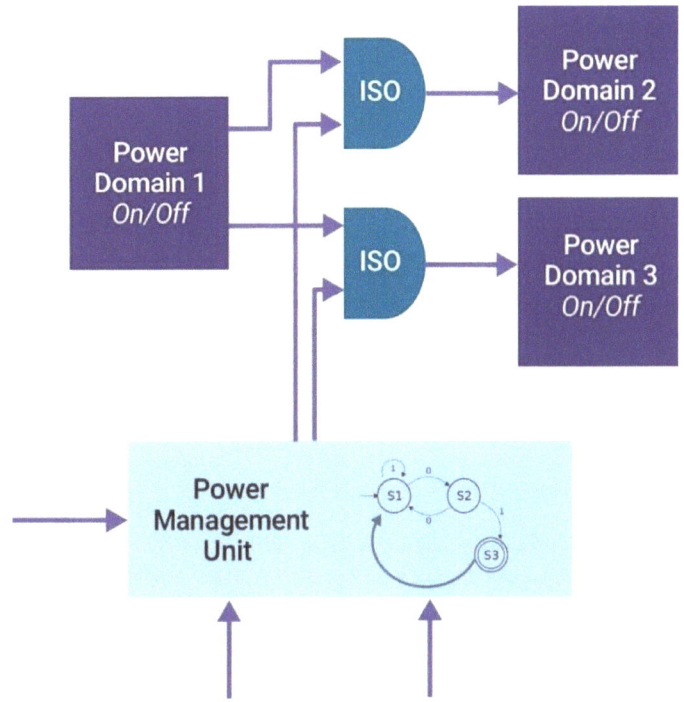

Formally verifying isolation controls based on power state

But think a little more about the isolation case – under what circumstances should isolation be enabled? This does depend on the use-case, so cannot be verified statically. Commonly verification teams further enhance their simulation/emulation testbenches to check these cases, which provides some level of increased confidence in coverage. But there's an obvious limitation in that approach which goes back to the reason formal methods first became popular. Creating, running and debugging a huge number of test cases to verify the correct behavior of the design *without power considerations* is already a huge task. Now imagine having to repeat that analysis across each of the possible power configurations for a design.

Talk about combinational explosion - this would be completely impractical.

Formal verification apps can help here. These look at power-state switching expressions to determine if there may be conflicts, providing you with confidence that, at least to this extent, all possible behaviors have been proven correct.

Architectural formal verification

Some problems cry out for formal verification because the number of cases that have to be considered is so high. But the system is far too big to fit in a formal proof and there are no obvious candidates to abstract to reduce the proof size to a reasonable level. The next logical step in these cases is to abstract almost everything! One good example can be found in cache-coherence checking where proofs have to span between multiple compute cores and their respective caches.

Multi-core architectures have become popular in SoCs for many reasons. These bring with them a well-known and still-challenging problem. To optimize performance, cores depend on local cache memories which provide faster local access to data than would be possible through main memory, while still syncing with main memory as needed.

However, those cores still need to assume that they are dealing with one logical memory model – the main memory. Cache memories are a hardware trick to speed up access, but they cannot break that logical assumption. This gets tricky when two or more cores are working through their respective caches with the same memory address, say address X. If core A updates the value at address X in its cache memory, then core B reads the value at address X in its cache memory, the value core B reads will be wrong. It ought to get the updated value but is unaware of the core A update so instead reads the outdated value in its own cache.

Cache-coherent architectures have clever ways to deal with these cases, which essentially come down to snooping on or otherwise being aware of addresses that have been changed by other cores. Any attempt to read such an address forces an update either locally or through main memory to ensure that all cores continue to see a consistent (coherent) logical view of the memory. As you might imagine, making this work is not simple when still trying to preserve most of the performance advantages of local caches. Proving correctness of operation in all cases, considering all cores and caches, is a complex control problem which is perfect for formal except for the fact that the whole system is too big.

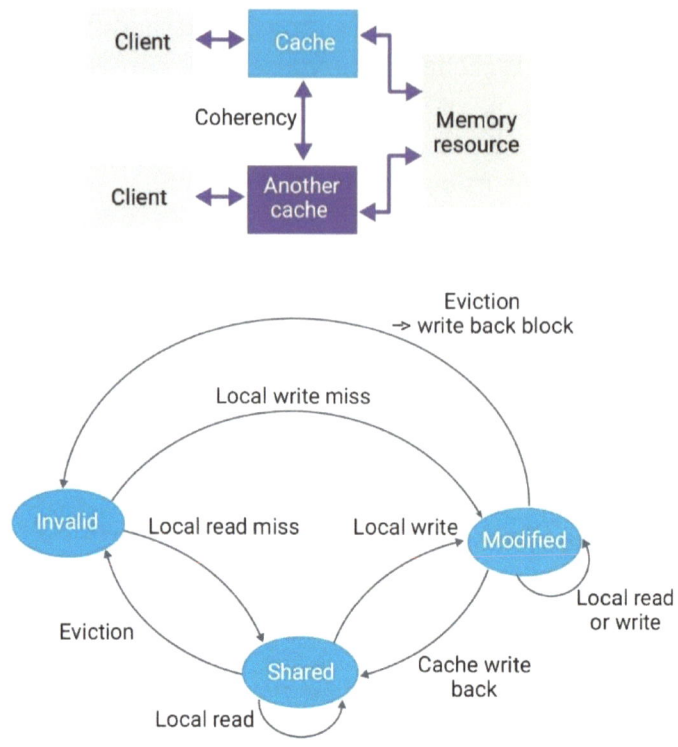

(Simplified) state diagram for cache coherence behavior

In an architectural formal analysis, you replace each of the endpoint IPs with a manually constructed FSM, modeling just those control behaviors that you consider important to the proof. There are generally three steps to this process: build each FSM starting from the architectural specification of the associated block (because it is more rigorous to check against what the architect wanted, rather than what the RTL team built). Then you run the formal proof on the abstracted system. Finally, you will validate each abstracted FSM against the full RTL for the corresponding block. This is driven by the assertions and constraints you added to the abstracted model, now used as input to the RTL model consistency check. This last step can be accomplished though simulation or formal verification methods.

Security verification

Security is a moving target; threats continue to evolve; therefore, defenses must also evolve. Unfortunately, security is an area where even 99% coverage isn't good enough. Hacks rely especially on rare and obscure weaknesses, so verification has to be as close to perfect as possible. Which means that formal methods are the only acceptable way to signoff.

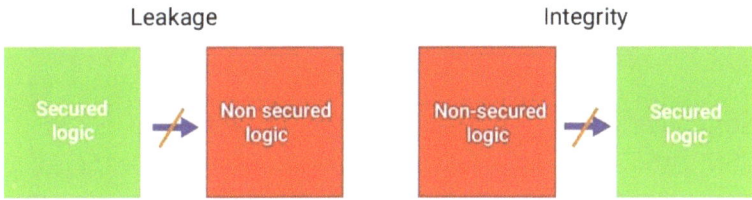

Checking for data leakage or data integrity compromise between secure and non-secure domains

The scope of what might possibly need to be checked can be rather large, including defenses against embedded hardware Trojans[74] and side-channel attacks[75]. However, the great majority of commercial activity is centered on attacks where software in an insecure domain tries to get access to data in a secure domain, by exploiting design or architecture weaknesses. As we have seen recently with the Meltdown and Spectre bugs[76], hardware is not

---

[74] https://en.wikipedia.org/wiki/Hardware_Trojan
[75] https://en.wikipedia.org/wiki/Side-channel_attack
[76] https://googleprojectzero.blogspot.com/

immune from security problems, despite all the advanced techniques that are already being used to limit attacks.

Given the range of security techniques that can be applied, and the intentionally limited scope of secure areas in many cases (to reduce the attack surface), this domain is a natural for app support. An app in this area should, for example, provide point-to-point checks to verify that data cannot leak from critical secure domains (such as cryptographic key-stores) to non-secure domains. Apps along these lines are already available; you should expect to see continuing development and research[77] in this area based on growing concern and awareness at all levels of society.

Safety verification

Safety verification, particularly functional safety verification, is another domain which is very hot, especially around automotive applications, but here the role of formal methods is a little different. A very important aspect of safety verification is in proving that static or transient faults in certain critical parts of the design will either not affect safe operation or will be appropriately mitigated by safety mechanisms designed to manage such failures.

---

[77] https://people.csail.mit.edu/nickolai/papers/chong-nsf-sfm.pdf

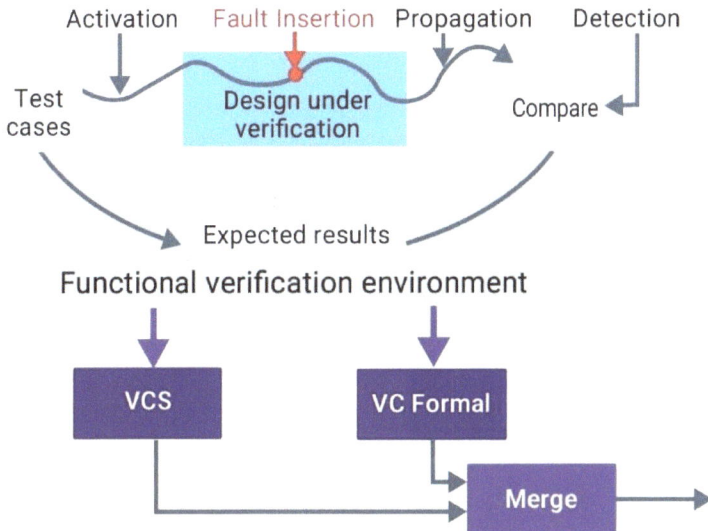

Using formal to filter fault that should be verified in safety checks

A common technique to perform this testing is to use mechanisms to inject faults, then determine if they will be detected at critical outputs. This works very well but can become expensive in simulation when faulting every node that might be critical, even when using clever incremental analysis techniques. Formal methods can be used to filter faults before simulation or after simulation, finding cases that cannot be observed or controlled, which should greatly reduce analysis and debug time[78][79].

Datapath Validation

We mentioned earlier that standard formal verification methods struggle with datapath components. A particular

---

[78] Automotive Safety and Security in a Verification Continuum Context
[79] How Formal Reduces Fault Analysis for ISO 26262

problem in arithmetic functions is the need to do bit-level analysis on word-level values, which quickly leads to explosion in the corresponding state-graph[80]. But we use datapath functions everywhere – in CPUs, GPUs, DSPs, cryptography, neural nets, GPS location and many other places (commonly these functions are add, subtract, multiply, divide, square-root on integer or floating-point values). Since all of these functions have well-defined specifications, they should be a natural fit for formal. In fact, the bigger companies are already using advanced formal methods such as theorem-proving to verify these designs.

Some of these methods are starting to appear in commercial tools, although they are not typically as well-known as the more mainstream formal methods. One example is the Synopsys Hector[81] product, optimized to validate high-level properties / equivalence for functions like datapath operators. You should expect solutions like this to play a more prominent role in formal verification platforms in the near future.

---

[80] Formal Hardware Verification: Methods and Systems in Comparison
[81] Formal Verification and Validation of High-Level Optimizations of Arithmetic Datapath Blocks

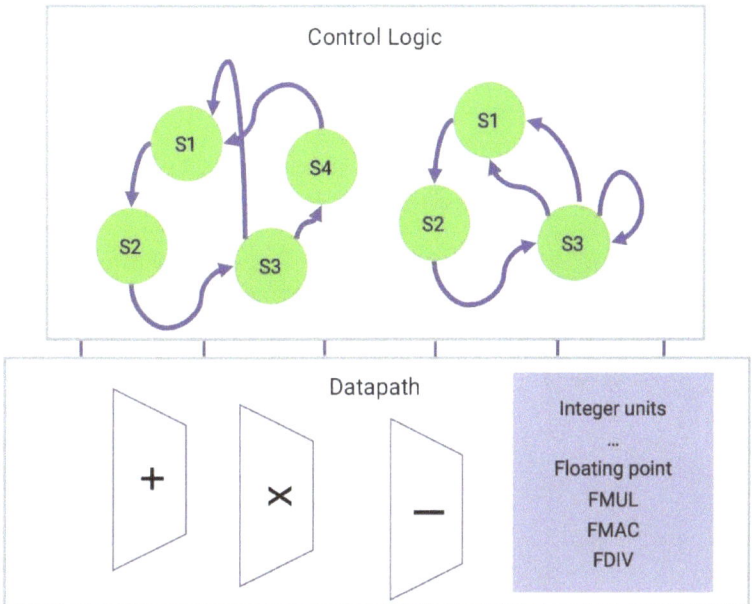
Datapath verification

Technology advances
Formal tools are always advancing – improving performance and heuristics to extend the size of problems that can be handled, improving parallelism to more effectively leverage scalable analysis in the cloud (a popular advance these days in many EDA tools) and adding further improvements in usability. We'll mention here a few examples of special interest.

Machine learning
A major area of development is in use of machine learning (ML) to simplify the use of formal methods. ML is hot in almost every domain today because it can capture and apply human skills not easily captured in algorithms (such as recognition, understanding and decision-making based on incomplete data) which are difficult or impractical for

humans to reasonably process. One example presented in a FMCAD Tutorial in 2016[82] simplifies debug through an intelligent interface.

> Describe your problem: I have a failure in assertion IO/A1. The values of the Ack signal seems to be invalid in the displayed waveform.
>
> Machine: Hi
>
> Human: Hello
>
> Machine: Could you please let me know what is your run-time session?
>
> Human: 200451
>
> Machine: Ok, I can access your FV session. What is the signal name?
>
> Human: IO/Ack
>
> Machine: Please specify the time at which the Ack signal is invalid.
>
> Machine: What is the value you expect?
>
> Human: At time t = 105, I expect the value to be 1.
>
> Machine: Can you specify why this is expected?
>
> Human: My constraints imply that a request should be successfully accepted. IO/REQ is high.
>
> Machine: Upon analysis of your design and constraints, Ack can be 1, if reset is 1 in your constraints and IO/REQ stays high from time 25 to 55. Would you like me to modify your constraints and re-run the design?
>
> Human: Yes, thanks!

A machine-learning bot conversation during formal debug

82

http://www.cs.utexas.edu/users/hunt/FMCAD/FMCAD16/slides/tutorial1.pdf

Modern debuggers for SoC verification are immensely capable but they depend on a wide range of modes and options to offer the features needed by specialists in each verification domain. Capabilities to support specific debug task are in the tool but learning how to use those features effectively can take a while. There is another option – in effect an expert chatbot supporting natural language processing (NLP). You can tell the bot in natural language what you want to do. It can then analyze/search and suggest a next step (again in natural language).

This conversation can proceed through multiple steps through a debug cycle until you get to a desired goal; possibly from finding a counter-example, to finding the problem was over-constrained, to correcting that problem and re-running, to finding a different counter-example, to finding a root-cause problem. You can still look at waveforms, but you skip all the "set this mode, set this option, pick this menu, …" stuff. This doesn't just make debugging easier, it also accelerates time to closure.

Advances in tools and methods

This is an area where vendors are naturally unwilling to share details, but we are free to speculate based on directions in academia and obvious needs around present methods.

There are multiple directions that are being taken in proof engines. One particularly interesting example is in use of induction-based methods. Induction is a mathematical technique of great antiquity, used to prove, in a finite number of steps, statements which may run through an infinite range of possibilities.

What makes this kind of proof appealing is that, at least where applicable, it has the potential to provide unbounded proofs. One obvious application is in proving the correctness of implementations of datapath elements (a multiplier for example).

> 1. You assume that a statement holds true for some number N
> 2. If you can then prove it must therefore also hold true for N+1
> 3. And you can separately show that it does indeed hold true for say N=0
>
> Then it must hold true for all possible N: from #3 it holds for N=0, from #1 and #2 it must also hold true for N=1, and by repeating it must hold for N=2, N=3 and so on forever.

Example of a simple induction

Induction also has relevance to abstraction[83], another area where tools continue to advance, particularly when it comes to automated abstraction. Ideally, you'd like a formal tool to figure out all the structures in your design which the tool knows will cause a proof to blow up, determine an appropriate abstraction for each of these functions and automatically (or in some guided manner) generate the command to replace those functions with the corresponding abstractions.

---

[83] http://prod.sandia.gov/techlib/access-control.cgi/2014/1420533.pdf

But designers have unbounded ingenuity in how they build functions, so automatic detection isn't easy, nor is automatic generation of an abstraction which will faithfully model the behavior of that function. So tool developers continue to advance the bounds of what they can detect and replace automatically, or if needed with a little guidance from a verification engineer.

What we discussed in this section has been only a sampler of R&D directions in formal verification. You may want to check out our suggested reading on *Deeper Background* for more on this topic.

The outlook for formal verification

There was a time when formal verification was looked on as a niche technology for specialized problems, but that view is now rare. Large systems and semiconductor companies now look on formal as a first-class component in any credible verification strategy, not as a way to replace other verification technologies but as an essential complement to those technologies. EDA companies have recognized this trend and are investing the same level of attention in their formal products that they invest in simulation-based methods. This is clearly signaled in the proliferation of easy-to-use apps and continuing expansion in problem sizes that can be addressed, along with very active research in directions like ML-assisted debug.

Much of what drives these advances is making formal verification more accessible to users who are not expert in formal methods, people like you who are expert in many other domains, who have a job to do, and are quite willing to embrace formal as a tool, but not necessarily as a career choice.

At a company level, pressure to use formal methods will always increase. As designs become more complex, more multi-core and more security- and safety-sensitive, formal proofs are often the only way to get to satisfactory closure. Greatly improved usability has effectively eliminated the barriers to adoption, since you can now ramp up formal skills along a gentle slope if needed.

And on a personal level, there is no question that adding proven skills in formal verification will greatly enhance your marketability[84]. Product teams are now looking for verification engineers well-versed in all or at least most verification techniques. Adding formal skills to your resume is a pretty obvious plus, especially since you no longer need an advanced degree to get there!

---

[84] If you doubt this, check out searches for formal verification engineers at companies like Google, Facebook and Apple

If you want to go deeper

There are many excellent and more detailed sources of information on formal verification. We have selected just a few we think you might want to follow, either based on continuing our theme of a high-level overview without getting too technical or, in the deeper background selection, information on the beginnings of this field, current state and where it is headed.

More detail for beginners

Erik Seligman: Formal Verification: An Essential Toolkit for Modern VLSI Design
Douglas Perry and Harry Foster: Applied Formal Verification: For Digital Circuit Design
Ashish Darbari and Iain Singleton: Industrial Strength Formal Using Abstractions
And, of course, training from the vendors who supply (or may, if you buy) your formal tools

Deeper background

Ed Clarke: Model Checking (MIT Press)
Rolf Dreschler: Formal System Verification: State-of the-Art and Future Trends
Ken McMillan: Symbolic Model Checking
Malay Ganai: SAT-Based Scalable Formal Verification Solutions

Glossary of formal terms

Like all of us, formal experts love their jargon and don't always understand that the rest of us may be confused by their specialized language. Fortunately, there aren't too many of these special terms:

Term	Informal definition
App	A pre-packaged application to make user involvement in checking some specific characteristic of a design much simpler than would be required through custom property checking
Assert	A requirement on logical behavior which can be checked in verification
Assume	A SystemVerilog constraint – see Constraint
Assume-guarantee	A way to simplify proof problems is to break the circuit into smaller pieces, say an upstream piece of logic and a downstream piece. You first constrain the inputs to the downstream piece (assume a certain behavior) and prove that piece functions as expected, then you use those assumptions as assertions on the outputs of the upstream piece and prove (guarantee) that those assumptions used in the downstream proof are valid.
BDD	Binary decision diagrams – a data-structure used in certain proof-

	engines such as for Symbolic Model Checking.
BMC	Bounded model checking – a type of formal engine which checks a property against the circuit in a breadth-first approach until either a counter-example is found, or a specified depth is reached.
Bounded proof	A case where no proof or counter-examples were found out to a specified proof depth, where checking stopped. The proof is bounded because there is no guarantee that counter-examples do not exist beyond that depth.
CEX	See counter-example
Constraint	A property which limits behavior of some set of signals in the circuit during proving. This could be as simple as fixing a signal value but can be as complex as a checking property. One example would be to define a one-hot constraint on a set of inputs.
Counter-example	An example of an execution path (generally presented as waveforms) which demonstrates that the property that you want to prove is clearly false.
Inconclusive	The result of a formal proof is inconclusive, if the proof cannot be

	completed or counter-example cannot be found within the specified time or memory bound by a given formal method, and therefore you can't conclude the correctness or incorrectness of a property. In the case of BMCs, the proof may terminate at a finite depth without finding a failure.
Model-checking	A formal technique which determines if a defined property or specification holds true for a given design or model.
Proof depth	One definition is the cycle bound reached on an inconclusive property result.
SAT	A type of formal engine which looks for a set of variable assignments which will satisfy the disproof of an assertion. This approach can be very fast since it isn't attempting to globally prove the truth of an assertion. It will stop as soon as a counter-example is found, if one exists (within the assigned proof depth).
Sequential depth	The number of clock-cycles required from the start of a proof to reach a certain goal – which might be testing an assertion or reaching a coverage property for example.

State space	The graph of all possible states and transitions in a design.
State space diameter (or radius)	The minimum number of cycles required to reach the farthest reachable state from the starting state.
SVA	SystemVerilog Assertions – the standard format used to express properties, assertions and constraints.
Vacuous proof	When testing a proposition "X implies Y", if X is false then, by a peculiarity of logic, the proposition is true. Of course, this is meaningless, which is why it is called vacuous. In formal verification this can happen with "if X then Y" assertions. If X is never exercised in proving (perhaps because of an over-constraint), the assertion will be reported as vacuously proved.
Witness	In the case of a property that holds true, a witness is one example of a path which demonstrates that the property is true.

Acknowledgements

While three of us are listed as authors, this book would not have been possible without the hard work, dedication and support of a much larger team.

Firstly, we owe a huge debt of gratitude to our Synopsys marketing leads (Sonia Montgomery and Kiran Vittal) who supported us in all aspects of the book. They were the engines that kept us moving forward over many months: keeping us on track, taking care of graphics, searching for background material and arranging meetings. They simplified the process tremendously, allowing us to simply focus on the content.

Special thanks are due to Jim Greene from the Samsung Austin Research Center, for finding time in his busy schedule to share with us his experience in formal verification through a compelling foreword. His experience and real-world application provide an encouraging example of how far it is possible to take formal verification with the right planning, investment and management alignment.

We want to thank our Synopsys Verification leaders, Michael Sanie and Mo Movahed, who had the vision and faith in us to put together this project. Without their funding and support this book would not exist. We also leaned heavily on the knowledge and experience of the VC Formal R&D and AE teams. Their insights added practical wisdom and background to this work.

We would especially like to thank our main Synopsys reviewers Pratik Mahajan and Ravindra Aneja for giving us valuable feedback on very short notice (sorry guys).

Finally, thanks are due to our families for their infinite patience in supporting us over the many long hours, evenings and weekends we consumed instead of spending time with them.

About the Authors

Bernard Murphy – SemiWiki

Bernard Murphy is a part-time blogger and author with SemiWiki, serves on the board of Mother Lode Wildlife Care in the California Gold Country and admits that he's never had so much fun. Earlier, he held down a real job as CTO at Atrenta. Earlier still, he held technical contributor, management, sales and marketing roles variously at Cadence, National Semiconductor, Fairchild and Harris Semiconductor. In his re-invention as a writer, Bernard has published over 300 blogs between SemiWiki and EETimes and has also released the book "SoC Emulation: Bursting into its prime", co-authored with Mentor Graphics. He received his BA in Physics and D. Phil in Nuclear Physics from the University of Oxford.

Manish Pandey – Synopsys

Manish Pandey is a Fellow at Synopsys, and an Adjunct Professor at Carnegie Mellon University. He completed his PhD in Computer Science from Carnegie Mellon University and a B. Tech. in Computer Science from the Indian Institute of Technology Kharagpur. He currently leads the R&D teams for formal and static technologies, and machine learning at Synopsys. He previously led the development of several static and formal verification technologies at Verplex and Cadence which are in widespread use in the industry. Manish has been the recipient of the IEEE Transaction in CAD Outstanding Young author award and holds over two dozen patents and refereed publications.

Sean Safarpour – Synopsys

Sean Safarpour is the Director of Application Engineering at Synopsys, where his team of specialists support the development and deployment of products such as VC Formal, Hector and Assertion IPs. He works closely with customers and R&D to solve their current verification challenges as well as to define and realize the next generation of formal applications. Prior to Synopsys, Sean was Director of R&D at Atrenta focused on new technology, and VP of Engineering and CTO at Vennsa Technologies, a start-up focused on automated root-cause analysis using formal techniques. Sean received his PhD from the University of Toronto where he completed his thesis entitled "Formal Methods in Automated Design Debugging".

www.ingramcontent.com/pod-product-compliance
Lightning Source LLC
Chambersburg PA
CBHW040218220526
45473CB00001B/31